PHOENIX ZONES

Phoenix Zones

WHERE STRENGTH IS BORN AND RESILIENCE LIVES

Hope Ferdowsian, MD, MPH

THE UNIVERSITY OF CHICAGO PRESS   CHICAGO AND LONDON

The University of Chicago Press, Chicago 60637
The University of Chicago Press, Ltd., London

For more information, contact the University of Chicago Press,
1427 E. 60th St., Chicago, IL 60637.
Published 2018
Printed in the United States of America

27 26 25 24 23 22 21 20 19 18 1 2 3 4 5

ISBN-13: 978-0-226-47593-6 (cloth)
ISBN-13: 978-0-226-47609-4 (e-book)
DOI: https://doi.org/10.7208/chicago/9780226476094.001.0001

Library of Congress Cataloging-in-Publication Data

Names: Ferdowsian, Hope, author.
Title: Phoenix zones : where strength is born and resilience lives /
Hope Ferdowsian.
Description: Chicago : The University of Chicago Press, 2018. |
Includes bibliographical references and index.
Identifiers: LCCN 2017032243 | ISBN 9780226475936 (cloth : alk. paper) |
ISBN 9780226476094 (e-book)
Subjects: LCSH: Psychic trauma—Case studies. | Victims of violent crimes—
Case studies. | Animal welfare—Case studies. | Psychic trauma. | Victims
of violent crimes—Services for. | Healing. | Compassion.
Classification: LCC BF175.5.P75 F47 2018 | DDC 362.19685/21—dc23
LC record available at https://lccn.loc.gov/2017032243

♾ This paper meets the requirements of ANSI/NISO Z39.48–1992
(Permanence of Paper).

To

Victoria, PD, Brigit, Maurice, Abi,

and too many others,

with a promise to do better

for so many like you

CONTENTS

PART THREE: THE RISE:
BUILDING PHOENIX ZONES IN A CONFLICTED WORLD

INTRODUCTION

In a quiet New York City courtroom, a group of adults listened to a small child sitting in a hard wooden chair as she described the abuse that dominated her home life. She could be one of many children today, but it was 1874 and the girl's name was Mary Ellen. Her alleged abuser was a woman she called "mamma."[1]

At the time of her trial, Mary Ellen was ten years old and living in Manhattan with her two guardians. When she was only a baby, her birth father, a Union soldier, had died, leaving her birth mother impoverished and unable to support her. By the time Mary Ellen reached the age of two, the city had assumed responsibility for her. Within a few years, she was placed in the custody of a woman named Mary McCormack ("mamma") and her first husband, who subsequently died. McCormack remarried a man named Francis Connolly and, together, they were Mary Ellen's foster parents at the time of her abuse.

Maryetta (Etta) Angell Wheeler, a Methodist caseworker, looked for Mary Ellen at the Connollys' home after learning about her situation. There, Wheeler discovered a young girl who appeared to be half of her nine years at the time, sparsely dressed despite the cold weather and covered in clearly identifiable wounds. In addition to being starved and beaten, Mary Ellen had been confined alone in a small dark room and deprived of affection.

Wheeler was compelled by Mary Ellen's case, but she had difficulty finding someone who would help the child. At the time, much of the public thought parents had a right to treat their children however they desired; they were in a real sense property. Few laws or organizations, if any, protected children from physical abuse at the hands of their parents or guardians.

As Wheeler became more desperate in her search, she approached Henry Bergh, a man well known for his kindness and political connections. Several years earlier, he had founded the American Society for the Prevention of Cruelty to Animals (ASPCA). Wheeler visited his office after her niece advised her: "You are so troubled over that abused child, why not go to Mr. Bergh? She is a little animal, surely."[2] Though at first bothered by the idea, Wheeler realized she had no other options.

When Wheeler initially appealed to Bergh, she thought he was somewhat taken aback by her request. He told her the case might require a new legal approach. Before meeting Wheeler, Bergh had tried to intervene for another child, Emily Thompson, following a plea from another woman. From her window, the woman could see Emily being beaten outside in her yard. Though the guardian who abused Emily was found guilty, Emily was still sent back to live with her. Bergh was understandably frustrated by the outcome. Nonetheless, he asked Wheeler to send him some information about Mary Ellen, and he promised to review the case. In the meantime, he sent an investigator to Mary Ellen's home.

Immediately after hearing from Wheeler and his investigator, Bergh recruited ASPCA attorney Elbridge T. Gerry to present a petition to the court on Mary Ellen's behalf. The petition alleged that Mary Ellen was "unlawfully and illegally restrained of its liberty" and "frequently during each day, severely whipped, beaten, struck, and bruised." Noting that "the marks of said beatings and bruises will appear plainly visible upon the body and limbs of the child at the present time,"[3] they asked for a writ of habeas corpus—from the Latin, meaning "you (shall) have the body." It was a novel approach, suggesting that Mary Ellen had a legal right to bodily liberty and integrity—a right to be considered "someone," not "something."

After receiving the petition, Superior Court Judge Abraham Lawrence issued a writ of habeas corpus and a special warrant requiring Mary Ellen's removal from the Connolly home and her appearance in court. A police officer carried Mary Ellen into the courtroom. Law enforcement officers had wrapped her body in a carriage blanket since she had so few clothes. Later, Wheeler noted that "Her body was bruised, her face disfigured, and the woman, as if to make testimony sure against herself, had the day before struck the child with a pair of shears, cutting a gash through the left eyebrow and down the cheek, fortunately escaping the eye."[4] The scar would remain with Mary Ellen through adulthood.

Jacob Riis, a reporter, quoted Bergh at the time of the court proceedings: "The child is an animal. If there is no justice for it as a human being, it shall at least have the rights of the cur in the street. It shall not be abused." Riis wrote that he had witnessed the first chapter of children's rights "being written under warrant of that made for the dog."[5]

After a brief period of deliberation in the court, Mary Connolly was found guilty of assault and battery. It never became clear as to whether Mary Ellen's foster father, Francis Connolly, had participated in the abuse. But within three months of meeting Wheeler, Mary Ellen was free of the confinement, neglect, and abuse she had sustained. Unlike Emily Thompson, she was never returned to her foster home. Instead, she was temporarily placed in a home for girls and later with adolescents under the authority of the Department of Charities and Correction. Out of concern for Mary Ellen's well-being, and with Bergh's help, Wheeler appealed to the court for guardianship. As Wheeler continued her work in New York City, her mother and sister helped raise Mary Ellen in a bucolic, wooded area of upstate New York.

Despite her early circumstances, Mary Ellen appeared to thrive in life. In her twenties, she married a widower and had two children with him. She adopted another daughter who had been orphaned, and she helped raise her husband's three other children from his previous marriage. She lived to the age of ninety-two.

Mary Ellen's case personalized child abuse. It was the start of a new era acknowledging the rights of children.

The day after young Mary Ellen's trial, the *New York Times* reported on Bergh's participation in an article called "Inhuman Treatment of a Little Waif—Her Treatment—a Mystery to Be Cleared Up": "It appears from proceedings had in Supreme Court . . . yesterday, in the case of a child named Mary Ellen, that Mr. Bergh does not confine the humane impulses of his heart to smoothing the pathway of the brute creation toward the grave or elsewhere, but he embraces within the sphere of his kindly efforts the human species also."[6] Bergh told the court his assistance with the case was prompted by his duties as a humane citizen, and he made it clear that he was not acting in his capacity as president of the ASPCA. However, he promised to avail himself in the future to defend children from cruelties perpetuated against them.

Bergh was born to privilege. In his early adult life, he traveled around the globe in relative luxury. This period of time appeared to be the first in which he became acutely aware of the plight of animals. During his travels, he witnessed a bullfight in Spain, which ignited his concerns. His journal reflections on the moment revealed his outrage at the violence directed toward animals. He saw it as uncivilized and immoral.

Through his political connections, Bergh was appointed to a brief diplomatic post in Russia. One day before retiring from his position, he intervened when he came across a carriage horse being beaten. After his resignation, Bergh and his wife made a trip to England. There, he spent time with the Earl of Harrowby, the head of the Royal Society for the Prevention of Cruelty to Animals. Bergh was inspired and hoped to start a similar organization in New York. Soon after he returned home, he drafted a Declaration of the Rights of Animals and asked his influential friends to sign on, in order to acquire a state charter for the ASPCA. He obtained the charter in 1866, eight years before he met Mary Ellen. Soon after he established the organization, he helped pass the first bill against animal cruelty in New York State.

After securing the force of the law, Bergh patrolled the streets of New York on a daily basis. His initial efforts focused on saving horses from neglect and abuse, but he soon expanded his work to protect cows kept in dairies. There, they were malnourished and living in dark and dirty conditions, increasing the risk of disease transmission to other cows and to any humans who drank their milk. Throughout Bergh's career, he fought for animals of many species. He brought to court a case about chickens, charging that they had been plucked and boiled alive during food processing. Elsewhere, he helped develop alternatives to pigeon shooting and tried to stop foxhunting. Agents working with the society helped him break into dogfighting and cockfighting rings, where they began confiscating animals and arresting organizers and audience members. Bergh also opposed circus practices involving animals and, in another instance, helped save some turtles taken from the tropics and shipped to New York on a vessel. The turtles had been placed on their backs, with their fins pierced and tied together. After Bergh arrested the captain and crew, he used the written testimony of a zoologist to show in court how the turtles suffered. At the time, he was publicly ridiculed for his efforts. Years later, he commented that he thought the turtle incident was great publicity for the society and its mission.[7]

Soon, the ASPCA's reach spread across the United States. Bergh worked with other concerned citizens to combat violent activities, like dogfights, that often moved from city to city. Though notable individuals like Louisa May Alcott and Ralph Waldo Emerson were supportive of his efforts on behalf of animals, the media often attacked him. Cartoons portrayed him riding through the streets in his suit and top hat, and he was frequently called "The Great Meddler"—seemingly for both his animal and child protection work.

Even though Bergh was consumed by his endless work on behalf of animals, he continued to fight for the rights of children. Two years after Mary Ellen's case was heard before the court, he and attorney Elbridge Gerry, with the help of others, cofounded the New York Society for the Prevention of Cruelty to Children. In its first year, the organization investigated more than three hundred cases of child abuse. By 1876, a bill had been passed to prevent and punish wrongs committed against children. Though Bergh remained on the organization's board, he felt adamant that his two groups should be kept separate. He thought combining the two would limit the efficacy of both causes, particularly since children and animals were already marginalized, with so few protections.

Together for years, but through separate organizations, Bergh and Gerry worked to enhance the protection of children and animals. But their societies represented only two of many groups working to improve lives. Others were also suffering. Slavery had just officially ended in the United States, and many African Americans were still fighting for freedom. American women had not yet won the right to vote. Immigrants faced backlash in uncertain economic times, and they were often characterized as useless animals. Specific groups, based on their sexual orientation or national origin, were banned from entering the country. Even with the emergence of a society specific to the protection of children, many boys and girls of color would not be afforded justice for many years thereafter.

Today, the children's rights and animal rights movements operate in very separate spheres, much like many other social causes. Few advocates in either sphere would call a child an animal, though the comments Bergh made about Mary Ellen being an animal weren't meant to demean her but instead to underscore her vulnerability to being harmed. Today, some hesitate to even acknowledge that we as human beings belong to the animal kingdom, or to consider what our inclusion in the animal kingdom implies about our shared vulnerability with other animals. Concern for the mutual vulnerability of children and animals united Bergh and Gerry in their efforts. A central ethic and a large circle of compassion, which saw species boundaries as morally irrelevant, guided them.

Much of our refusal to accept the human-animal bond takes the form of what primatologist Frans de Waal has called "anthropodenial": a taboo against granting that animals have emotions like humans, or that humans have emotions like animals.[8] Despite centuries of scientific studies and logic showing that animals have internal lives and indeed suffer, the predominant ethic and treatment of animals lags behind gains in knowledge.

When he first penned the term "anthropodenial," de Waal was responding to the still-common use of the term "anthropomorphism." The word stems from the Greek and, loosely translated, means "human form." At the time of the word's origin, critics objected to literature that treated Zeus and other gods as if they were people. The concern was not with a comparison to animals, but with the seemingly arrogant comparison of humans to the gods. Today, the term "anthropomorphism" is sometimes used to criticize similarities made between people and animals. Regardless, says de Waal, "Modern biology leaves us no choice other than to conclude that we are animals. In terms of anatomy, physiology, and neurology we are really no more exceptional than, say, an elephant or a platypus is in its own way. Even such presumed hallmarks of humanity as warfare, politics, culture, morality, and language may not be completely unprecedented."[9] De Waal's assertions have been repeatedly tested and affirmed. Nonetheless, anthropodenial continues, perhaps because of our discomfort with how we treat the very animals we resemble.

Pains to minimize the complex inner lives of animals, to make them less than they are, call to mind many efforts used to subjugate some humans over the course of history. It removes them from the moral equation. They lose, and so do we.

Mary Ellen's nearly one-hundred-fifty-year-old story is like many stories I've heard as a human rights physician and as an advocate for animals. Bergh realized that Mary Ellen's experiences mirrored the grim sagas of the animals he cared about. But the story of how Bergh and Mary Ellen came together offers only a glimpse of the important connections between violence against people and animals.

Today, policymakers attempt to address violence in all forms, from that which occurs in the home to the street to military conflict zones. Economists, social engineers, psychologists, neuroscientists, and many others join them. Nevertheless, these problems persist. Some prominent intellectuals differ in opinion about whether overall levels of violence have decreased since the Second World War. Regardless, we are far from the zero mark in our quest to eliminate violence. For those caught up in violence, whose vulnerability to suffering is deepened on a daily basis, the argument over whether it is increasing or decreasing is merely academic. Their lives and turmoil suggest that we need better solutions that address its roots, including the link between human and animal suffering. This book contends that we cannot address one without addressing the other.

I grew up on a small farm in Oklahoma and knew as a child that I wanted to become a doctor. Early on, through my parents, I learned about human rights violations around the world. I became intrigued by international affairs in college, where I studied factors that lead to genocide, torture, and other human atrocities. Rarely, if ever, did my professors discuss comprehensive solutions to these problems. By the time I entered medical school, I had also become sensitized to the suffering of animals around the world. After seeing the ways animals are treated in society, I couldn't help but recall the animals I knew and loved as a child. It was then that I began to seriously consider the relationship between the plight of people and animals.

Today, as an internist and preventive medicine physician, I straddle the fields of medicine, public health, and ethics. I've been fortunate to pursue work that bridges my concern for people and animals. Internationally, I've worked throughout Africa, Micronesia, and other parts of the world. In the United States, I've worked with nonprofit organizations providing health care and advocacy for homeless, immigrant, and other marginalized populations in urban and rural areas. Over time, I have found it impossible to separate my reflections on patient care and the human rights and animal protection efforts in which I have been involved. But until fairly recently, I found it difficult to articulate the deepest links between human and animal well-being. That was, until I began to consider the importance of a central ethic and many of the principles by which Bergh lived.

Beginning in 2010, my colleagues and I received federal funding to explore ethical problems with the use of animals in research, as well as some ongoing challenges within human research. With the goal of reducing suffering, I brought together professionals from human and veterinary medicine, industry, philosophy and ethics, advocacy, and the policy realm to determine whether some human research protections could be extended to animals. Together, we studied how concepts traditionally reserved for human research ethics, such as respect for autonomous decision-making and consent, obligations to avoid harm, and the demands of justice, could be applied to questions about the use of animals in research. In the process, I began to realize the potential influence of ethical principles, and how commonly held moral values could be applied to critical decisions in medicine, research, and other areas of society.

As fundamental truths or propositions, principles serve as the foundation for values, behaviors, and logic.[10] Bioethical principles used within the context of medicine and research relate to broader moral norms within society—for example, respect for autonomy reflects our reverence for liberty, and attempts to avoid harm echo the importance of kindness and compassion in society. Over time, I began to question whether a broad framework of moral principles could also have some sort of *therapeutic* value—both for society and for individuals. I remained preoccupied with this question as I traveled and came upon a promising realization.

During the course of my work, I realized I was witnessing how people and animals can thrive after severe trauma—a transformation known in medicine as the "Phoenix Effect." Changes in biological mechanisms, brain structure, and clinical signs seen in people and animals help explain this phenomenon. As cutting-edge scientific discoveries reveal, animals can become traumatized in ways that are akin to human trauma. But they can also heal in similar ways, revealing a common scientific basis for resilience. The success stories of torture survivors seeking asylum, children like Mary Ellen with histories of abuse, animals healing in sanctuaries, and similarly wounded individuals hold noteworthy similarities. They point to important foundational values—principles—that can heal some of the most broken among us and cultivate even broader social progress. These ideals are found in what I call "Phoenix Zones."

This book begins by examining the relationship between violence directed at people and animals. It goes on to reveal how the core values found in Phoenix Zones can reduce vulnerability to suffering and foster resilience. These principles span from respect for freedom and sovereignty to love and justice to a regard for individual worth. Together, as physical and psychological needs, they form the biological foundation for the Phoenix Effect. The book concludes by probing how these pillars could help solve some of the biggest ethical, legal, and political challenges of our time—from escalating violence and terror to systematic cruelty and abuse.

Within this book are a combination of stories, facts, and figures. I have included stories that are both public and private. I've altered details to protect the identities of some involved, and I've asked some individuals I have written about to review the material for accuracy before including it. Through my work, I have had the benefit of learning from the hardships others have endured. I cannot tell their stories exactly as they could, but their experiences reveal compelling insights that deserve to be heard. I've taken a similar approach in telling the stories of animals and the people who know them best. Though animals cannot communicate to us their experiences in ways we fully comprehend, it is important that we know them for *who* they are, not *what* we wish them to be.

I have taken some liberties with the use of language. I use terms like "people" and "humans" interchangeably, while varying the use of phrases like "animals," "nonhuman animals," and "other animals." I realize there is ongoing debate about the use of these terms, particularly in light of evolving legal interpretations of the word "person." All expressions used to describe nonhuman animals seem both imperfect and insufficient, especially in characterizing their relationships with humans. Although ongoing deliberations about the best ways to refer to other animals are important, I hope the stories included in this book speak for themselves.

Finally, when I write about violence, I am speaking of it in its broad sense: intentional physical, psychological, or emotional harm. Sexual violence is but one example: as a violation of bodily sovereignty, it can cause emotional or psychological harm, beyond physical manifestations. Nor does violence necessarily imply criminality, or legality imply nonviolence. There are still many legal forms of violence in the United States and around the world.

During the period of time in which I wrote this book, there was an international rise in nationalism and racial and ethnic discrimination, bolstered by fear and uncertainty. Global politics were marred by the exploitation of bigotry, and of what divides us. Connections between fear, uncertainty, and division are not coincidental. As neuroscientists have shown, fear and intolerance for uncertainty can widen the empathy gap. But scientists have also shown how a consistent moral framework based on clear principles can override fear and prejudice and encourage unity and progress.

We are at a crossroads, or perhaps many crossroads. The timing is urgent. As Thomas Friedman has pointed out, we live in an age of accelerations driven by three forces: Moore's law (technology); the Market (globalization); and Mother Nature (climate change and what he calls biodiversity loss).[11] But urgency also presents opportunity. Within the fields of medicine and public health, one of the best times to build a new health care system is right after a crisis.

Now, we have a chance to examine what bonds us, instead of what divides us. To do so, we need to look honestly at our common propensity for vulnerability and resilience. If we ignore what unites us all, including our animality, we will miss an incredible opportunity.

More and more of us are looking for sanctuary—a respite from the torment in the world. Exhausted and overwhelmed by news about how we are divided, and by the constant influx of disturbing images and stories, we are hungry for concrete solutions. We battle with how to find hope amid despair.

This book aims to ease that exhaustion and lay out a realistic and optimistic call to action without shying away from difficult subject matter.

Writing about the courageous individuals in this book has inspired and energized me. My wish is that you will be similarly encouraged after reading their stories. Together, I hope we can share and expand a new narrative, one that acknowledges the pain in the world but also sets a more resilient path, a broader movement dedicated to improving the lives of all people and animals.

Much of the book and the stories in it are pinned on hope, sometimes the only medicine available. But true hope implies a willingness to take action, to change for the better. Like Henry Bergh, I believe we can.

PART ONE

The Clues

FINDING HOPE AMID DESPAIR

1 : THE PHOENIX EFFECT

FROM OPPRESSION AND VULNERABILITY TO STRENGTH AND RESILIENCE

The phoenix hope can wing her flight thro' the vast deserts of the skies, and still defying fortune's spite, revive, and from her ashes rise.

— MIGUEL DE CERVANTES SAAVEDRA, *Don Quixote*

Humidity and emotion filled the air. The expression on each face seemed to move in slow motion from fear to disgust to shock to relief. As my colleague Jacob spoke with passion, in a rural area of Kenya, a group of doctors, nurses, police officers, attorneys, and judges listened closely. Looking on, I sat in the back of the stark room littered with flip chart paper taped to coarse white walls. We were nearing the end of an exhausting week of training health and legal professionals in the community to take care of sexual violence survivors. We had been asked to focus the training sessions on caring for children; clinicians in the region had noted an increase in the number of young children and adolescents being sexually assaulted. Jacob was sharing his own experiences as a police officer working in the northeastern part of Kenya, near the Somalia border.

Standing in a white button-down shirt and dark trousers, Jacob described how he was patrolling on foot in an area near a refugee camp when he heard high-pitched, frenzied cries. It was near dusk, heightening his anxiety. He followed the screams to a small concrete building and came upon a young girl being assaulted by an older man. Her shaking body was wrapped in a torn stretch of cloth tied over her shoulder and draped around her waist: traditional dress known as a *guntiino*. She looked up at Jacob with wide eyes. She was only twelve years old, but her clear, round face made her appear even younger. The girl, Aiyana, had been forced into marriage with an older man—her attacker.

Jacob arrested the man, though he correctly predicted he would face death threats because of his actions. After removing Aiyana from the crime scene, Jacob tried to find her family. He discovered her father had taken her out of school and sold her to the older man. He didn't uncover exactly why Aiyana's father had traded her life and body for money, but poverty often plays a major role in parents' decisions to marry off their underage girls. It leaves one less mouth to feed, and in some places it's still tradition for the bride's family to receive a dowry: compensation for taking a daughter.

Though "child marriage" is still legal in some countries around the world, it is illegal in Kenya, where the minimum age of marriage is set at eighteen years. Adopted in 2014, the relatively new legislation applies to all forms of marriage, including religious and customary unions. Nonetheless, about one-quarter of all girls in Kenya marry before the age of eighteen. This statistic is slightly lower than the global average. About one-third of all girls in the developing world are married before their eighteenth birthday. In many of these cases, girls are forced into marriage with an older man, which commonly results in physical and sexual violence. According to a survey conducted by the International Center for Research on Women, girls who marry before the age of eighteen years are twice as likely to report being beaten, slapped, or threatened by their husbands as girls who marry later. When asked about the prior six months of their lives, they are about three times as likely to report being forced into sex.[1]

Many young girls coerced into marriage are also forced to bear children, a particular danger since their bodies aren't yet fully developed. Many don't have access to good medical care. Some girls are left to care for children on their own. In some cases, their husbands abandon them, leaving them as single mothers with no income and few job choices. Since they are pulled out of school at such an early age, their financial future is often bleak, perpetuating a relentless cycle of unfulfilled potential.

In theory, girls are protected from child marriage through international law. Both the United Nations' Universal Declaration of Human Rights and the Convention on the Rights of the Child call it a violation of children's rights. However, these treaties are poorly enforced. The

United States has not yet even ratified the Convention on the Rights of the Child. Fortunately, there are many groups and organizations working to end child marriage, including Girls Not Brides, a global partnership of hundreds of civil society organizations committed to enabling girls to fulfill their potential.

Jacob knew the statistics when he saved Aiyana. He had seen too many children shattered by sexual abuse. When Jacob learned about the intentions of Aiyana's father, he realized it was a difficult situation. Aiyana had nowhere safe to go, and Jacob feared she would be returned to a dangerous place if he didn't find an alternative. After talking with Aiyana about what she wanted to do, and working with social services, Jacob helped her find a new home and school, hundreds of miles from the trauma she suffered.

Though Aiyana's story ended better than it could have, many stories like hers do not. My colleagues and I are constantly reminded of this stark reality through our work to end impunity for sexual violence in areas marked by conflict and unrest. By the end of that week in Kenya, we were left with a mix of sadness, rage, and hope. And a fundamental question: How can we reduce vulnerability and prevent suffering, especially for those like Aiyana who cannot always protect themselves? As I prepared to leave Kenya, I had no idea how a vulnerable dog I met on a congested Nairobi highway would test these reflections: one of a series of powerful experiences that led me to think more holistically about how we can turn vulnerability into strength and resilience.

LOVE

As we drove back to Nairobi, I stared out the window and saw shadows of people lining the side of the road near shops made of red and brown tin. We were in the back of a white utility van moving at less than five miles per hour on the highway entering Nairobi. Though it was dark and cool, the air was thick. The smell of charcoal lingered, mixed with exhaust fumes from dented cars and trucks with missing parts. Oncoming vehicles provided the only source of illumination. Looking up from the windows smeared with dust and grime, I saw only darkness.

Our five-hour trip had taken us from rural Kenya, which was scattered with people and animals making their journeys by foot, to a

Nairobi highway crowded with cars, buses, and trucks. Night had just fallen as we drove into the city, and traffic had slowed to a near stop. A car had struck someone or something and continued on, as other cars maneuvered around the obstacle. I looked down and saw wide eyes looking up from the road. I looked at my colleague and saw tears welling up in her eyes. She confirmed what I had witnessed, a live dog lying still.

I insisted we stop. Another of my colleagues thought I had mistaken the dog for a person. She pointed out that the dog was not human, but "just a dog." Fortunately, our driver turned around, but only to let another colleague out to find a restroom. I thought of Jacob's courage in a far more dangerous situation, and I didn't hesitate. I jumped out of the van and ran toward a security guard in a blue uniform who had pulled the dog to a grassy median. I asked him to take me to him. The security guard put out his hand to stop oncoming traffic, and it slowly came to a reluctant halt.

When I reached him, the dog was whimpering in pain. He looked up at me with his huge brown eyes. His body was riddled with open wounds and scabs. He had old scars and scratches on his face, a torn blond floppy ear, a bitten nose, and severe fractures in his front and hind legs. Fleas and ticks covered his body. He had been neglected and abused. Michele, my colleague who had requested a bathroom break, helped me pick him up with some clothes I had grabbed from my bag. Since I wasn't sure if he would bite out of fear, I took some old scrubs to tie as a muzzle if needed. It wasn't necessary. He easily let me pick him up, and I held him close as he trembled.

A severed electric cord was wrapped around his neck; he had chewed his way free. Michele and I carried him to our van and, despite earlier objections, our colleagues had already arranged for a veterinarian to meet us at his clinic.

On the way, the dog shook uncontrollably in my arms. I tried to keep him from being jostled around by the potholes cluttering the back roads of Nairobi, so his broken bones wouldn't cause him more pain.

We reached the veterinarian's office in fifteen minutes. It was a Friday night, so the city was busy, but we drove into a quiet, dark driveway in a dirt alleyway. We walked into the clinic and heard the sounds of dogs barking in the back. A technician greeted us, and we waited another thirty minutes for the veterinarian to arrive. As we surveyed the dog's injuries together, we realized he suffered from multiple serious wounds and shock. He was unlikely to pull through surgery. His eyes drifted open and closed as we held him, and I tried to figure out what he wanted. After asking the veterinarian what types of medications would be used, we made the difficult decision to euthanize him. We wanted to ensure that he wouldn't suffer or experience a prolonged death. He lay in my arms while Michele and I gently stroked him and, through soft, soothing words and tears, we whispered that we loved him. He died quickly. Michele told me, "At least he knew love before he died." The veterinarian asked me to name him; I named him Love.

DOC

When I was about nine years old, around the same time I decided I wanted to become a doctor, I crawled up on my mom's built-in bookshelves and reached for a shiny red book. I had seen my mother reading the book with sad eyes and a long, drawn frown on her face, and I was curious to know what it was about. I was intrigued by the title—*A Cry from the Heart.* Written by William Sears in the 1980s, the book details the persecution of a minority religious group in Iran. As I opened the book and turned its pages, I read about a child who had been burned to death because of his religion.[2] I couldn't believe something so horrible had happened to a child like me, and I wanted my friends to know, too. I'm not sure how it happened, but I used Sears's book for an oral book report in the fourth grade. I still think it's a testament to the courage of my mother and my fourth-grade teacher that I was allowed to speak about such a frightening incident to my nine- and ten-year-old classmates. No parent complained, and after I read my report to the class, my friends and I all returned to the luxury of our less solemn books and the playground. Today, that image of a child being burned to death still lingers in my mind.

Almost thirty years after I picked up that shiny red book—but before I had ever learned about Aiyana or Love—I walked toward the glass door of my clinic waiting room and saw a man with a slight build and graying hair. It was the middle of the week, toward the end of the day, after all the other patients had been seen. The man was the shape and color of my father and from the same part of the world, one I knew only through the stories my father had told me. Kind, bowed, and humble, the man came toward me in a hesitant rush. With both anticipation and trepidation, he reached out his hand and I took it. He cradled my hand in his own and then released it. After he followed me into a bright clinical examination room, we both took a seat. Though he was an established expert in a rare field of medicine, he was unassuming. Like so many torture survivors I've met, he exuded a calm, quiet dignity.

The clinic is where I practiced medicine as a volunteer physician for nine years. In many ways that clinic became my own sanctuary in a profession now driven by everything medicine isn't supposed to be. It's where I watched other doctors, nurses, lawyers, social workers, and front-desk personnel live and breathe the words "respect," "justice," "compassion," and "hope." It is where I learned to be the doctor I wanted to be.

As I grasped soon after we met, the man, whom I'll call "Doc," had been tortured by the same ruthless regime that killed the small boy in William Sears's book. After introducing ourselves, Doc told me how Iranian government officials targeted, detained, and tortured him. In the 1980s, he became politically active after his colleagues were executed for speaking out against the government. On a cool winter night, government agents came to his home and arrested him. He was terrified for his family and tried not to resist or infuriate the officers.

After he was arrested, Doc was taken to a prison, where he was interrogated and beaten. He was held down and restrained on what he described as a "torture bench." Among human political prisoners and detainees, confinement to overly small spaces, restraint, and multiple forms of physical injuries are common torture techniques. Victims end up in fixed, hyperextended, or other unnatural positions, leading to short- and long-term pain and injuries to ligaments, tendons, nerves, and blood vessels. When I hear survivors' stories, I am often reminded of a poem by Wisława Szymborska, who lived through the Germans' occupation of Poland and won the 1996 Nobel Prize for Literature. She writes in her poem "Tortures" that bones are breakable and joints are stretchable—and both facts are taken into account by torturers; "The body writhes, jerks, and tugs . . . bruises, swells, drools, and bleeds."[3]

While he was restrained, Doc was incessantly beaten on the soles of his feet with an electric cable, a common torture technique practiced all over the world. Doc was continually questioned and threatened with his life. In a subtle Farsi accent, he told me, "I closed my eyes to shield my emotional response to the beatings." He explained to me that if he lived, he didn't want to remember the faces of his torturers or the tools used to torture him.

Once he was released, Doc lived in fear. From 1988 to 1998, there were a series of murders and disappearances of Iranian dissident intellectuals who were critical of the Islamic Republic government. This series of murders has been referred to as the "chain murders of Iran," and many believe the killings occurred in response to reformists' attempts to open up Iran's cultural and political place in the world. Victims included journalists, doctors, and other citizens. After protesting the chain murders of Iran, Doc was arrested again. He noticed the torture methods had become more sophisticated. Parts of his body were covered to prevent the formation of certain scars—forensic evidence that, if documented, could be used against perpetrators. After weeks of repeated torture, he was placed in solitary confinement. As we sat together in the clinic room, he dropped his head to his hands and told me he felt "broken."

The use of torture to intimidate and break victims has a long history. The ancient Greeks and Romans were among the first, but not the last, to systematically use and rely on torture as a means toward domination and oppression. Much as it is today, torture was commissioned to extract the "truth" from slaves who were considered unreliable witnesses, even though many ancient Romans acknowledged the unreliability of information obtained under torture. By the third century, the ancients' slippery slope of torture had slid down class lines, rendering virtually no one immune to torture. And by the mid-thirteenth century, even court judges and leaders of the Christian Church sanctioned torture. Though activists had ushered in attempts to abolish torture in democratic nations by the eighteenth century, communist and fascist governments widely relied on torture to maintain control. Democratic nations were at times complicit, continuing to support torture perpetuated by authoritarian commands. By the middle of the twentieth century, under the Nazi regime, torture often took the form of medical experiments. American medical experts even colluded with the Nazis: as observers, in experimental design and execution, and by excusing torture.

Though torture occurred long before the Gestapo, Nazi atrocities motivated the international community to end it. Soon after the Nuremberg Trials began in 1948, the United Nations General Assembly adopted the Universal Declaration of Human Rights. It marked an international commitment to eliminate torture. Over the next half century, governments enacted agreements banning torture. By 1984 the United Nations had unanimously approved the legally binding Convention against Torture. Today, if survivors can demonstrate a credible fear of torture, they are eligible for protection in countries such as the United States. Doctors like me conduct forensic examinations for people seeking asylum to evaluate whether there are indications of torture that can be used in their court cases. My work in this area relates to my work in Kenya, where I teach other health professionals how to conduct forensic exams on sexual violence survivors so perpetrators can be prosecuted.

As I do for other people seeking asylum in the United States, I provided a pro bono clinical examination of Doc to determine whether there was objective medical and psychiatric evidence that he had been tortured. After listening to Doc's story, I measured his visible scars and carefully documented their location, texture, size, shape, and color. Far more time was needed to gradually unravel his psychological wounds, which are often more severe. Even though many of Doc's external wounds have healed, psychological wounds remain deep. His nightmares resurfaced years after he was tortured.

Months after I wrote a medical report substantiating his claims, I learned from Doc's attorney that he would not be forced to return to Iran. Today, Doc is an active member of the medical community. In his hospital, he mentors his colleagues. He has reunited with his family, and he continues to pursue interests outside medicine, including history and gardening. He has become a friend.

A few years after our first encounter, Doc and I met over iced tea. We spoke about politics, religion, and history. He shared a story about the first human rights charter, which ironically came from ancient Persia, around the same time the ancient Romans and Greeks started using torture to extract confessions. In 539 BCE, Cyrus the Great, the first king of ancient Persia, entered the old city of Babylon, freed the slaves, and established religious and racial equality. Around that time, a baked-clay cylinder with cuneiform script in the now-extinct Akkadian language was buried beneath some walls; it was excavated in 1879.[4] In 1971, this cylinder was described as the world's first human rights charter since it called for racial and religious equality and an end to slavery and other forms of oppression. The charter has been translated into all six official languages of the United Nations, and its provisions resemble the first four Articles of the Universal Declaration of Human Rights. Doc told me how our views about human rights are not new but rather ancient ideas that require safeguarding and expansion—not unlike ideas about animal rights proposed by the ancients. Doc sought these ageless ideals in a thriving, open society in the United States. And today, safe from harm and the risk of deportation, with the love of his family, the respect of his colleagues, and freedom and opportunity, he is healing, rising from the ashes, and helping others rise as well. He, too, has led me to think more broadly about how we can turn oppression into hope and vulnerability into strength.

Over the past decade, through my work as a human rights physician, I've met countless men, women, and children who have suffered through unimaginable trauma. Like Doc, many have fled unsafe, oppressive lands, seeking refuge. They have witnessed unimaginable atrocities. Today, there are more people displaced by war, conflict, and persecution than ever before. They are doctors, lawyers, teachers, and journalists. Others are mechanics, police officers, bankers, and business owners. They are also fathers, mothers, brothers, sisters, sons, and daughters. And, like Doc, many have been tortured.

Torture still occurs in the majority of countries around the world, often in places that are hidden from the public. Anyone can be a victim of torture, and many torture survivors never figure out why they were targeted. They have survived severe beatings, electric shocks, rape, mock executions, starvation, sleep deprivation, and other abuses. Their bodies and minds bear discernible and hidden scars—including chronic pain, broken bones, traumatic brain injury, terminal diseases, and mental health challenges. In many ways, torture takes advantage of the body's most basic needs—the need to breathe, eat, drink, sleep, move, and remain whole and intact. As long as our bodies can experience fear and pain, torturers exploit our innermost vulnerabilities. And the bodies and minds of animals—like Love—are similarly vulnerable to torture and other forms of violence.

GRACE

Currently, there are about half a million torture survivors living in the United States. Many of these immigrants have a desire to contribute to society in precisely the ways that made them targets of human rights abuses in their home countries. Just as Emma Lazarus's famous poem, "The New Colossus," reads on the pedestal of the Statue of Liberty, they are "yearning to breathe free."

I finally realized the meaning of these words while standing on the white marble steps of a federal court building outside the nation's capital. There I stood with a woman named Grace, who pulled me toward her and shouted in my ear: "I'm free! I'm free!" About one year prior, Grace had fled her home in Africa after being tortured. As a result of being raped, she contracted HIV. She arrived in the United States after a long and difficult journey.

On the day Grace gained freedom, I was at the courthouse to provide oral testimony on her behalf. I had already provided written testimony substantiating her claims, based on my forensic medical evaluation. Although each story is unique, there are some threads of commonality among torture survivors. In addition to her physically discernible scars, Grace suffered from posttraumatic stress disorder (PTSD). Psychological trauma is overpowering and pervasive. Frequently, survivors struggle with chronic psychiatric disorders, especially PTSD and depression. After I concluded that Grace demonstrated objective physical and psychological evidence of torture, her attorneys used my medical report in her legal case.

Outside the courthouse, Grace looked at me with a smile that reached her deep-set brown eyes. She still lived with a potentially lethal medical condition. She was starting her life over in a new country with little social support and the long-term physical and emotional ramifications of torture. But she *was* free. Her life would no longer be threatened because of her political beliefs, gender, or ethnic background. After being held captive, deprived of her most basic needs, and physically and sexually assaulted, she was finally granted asylum. Like Doc, she found sanctuary in the United States.

When I first met Grace, I learned she had a young son in the United States. Grace became pregnant when she was raped and tortured. She spoke of her son with complete love; she felt no animosity toward him. She insisted the love she shared with him helped her heal. From the instant I met her to the moment we said good-bye, she showed unwavering strength and valor, in spite of how she was exploited by her perpetrators. Like so many other women I've met in her position, she demonstrated a level of resilience that seemed implausible after what she lived through, but was very real.

THE PHOENIX EFFECT

The stories of Aiyana, Love, Doc, and Grace are deeply linked—through the oppression, vulnerability, suffering, and resilience of people and animals—connections that led to the conception of this book. I have slowly begun to see a meaningful pattern in these stories: one that includes hope amid violence and aggression.

When I returned to the United States from Kenya, after learning about Aiyana and meeting Love, I struggled to grasp how we could better protect children like Aiyana, or other vulnerable beings, like Love. At the same time, I reflected on my decade of experience working with survivors of torture and sexual violence, as well as my experiences in the area of animal protection. I couldn't separate these issues in my mind. I knew that animals, like people across the world, were subjected to multiple, compounded sources of neglect, cruelty, and abuse over the course of their lives. I felt a compelling obligation to understand the links between violence against vulnerable people and animals, to encourage change. I was desperate to find hope and answers.

At the time of Love's death, I saw no other options for him. However, I have since wondered if despair temporarily blinds us to answers that hope provides. Is it actually possible to build refuge for those like Aiyana, Love, and other beings shaken by tragedy—so they can not only survive but thrive, as Doc and Grace have? And if so, how?

In Grace and Doc, I began to recognize a phenomenon I had witnessed repeatedly in others I have cared for—people defying their seen and unseen injuries and thriving after indescribable trauma, much like the mythical Phoenix. In Chinese, Egyptian, Indian, and Greek mythology, the Phoenix is a magical bird who is cyclically reborn. She rises from the ashes of her previous form. Even after living through what can only be described as hell, some survivors rebound, recover, and heal. Through Grace, Doc, and many other survivors I've met, I've come to understand how it's possible for despair to metamorphose into hope, almost like a physical and emotional rebirth. In medicine, this transformation is called the Phoenix Effect. With time, I've begun to view survivors as spirited Phoenixes who ascend from the ashes, despite the odds against them. And over the years, I've realized how animals can also heal after severe pain and suffering—offering a model for recovery and hope for how we can all rise from the depths of suffering. These stories, and the science behind them, are also a metaphor for how society at large can turn oppression and vulnerability into hope and resilience.

INSPIRING HOPE THROUGH THE PHOENIX EFFECT

It's tough to turn toward suffering. In medicine, journalism, law enforcement, and other professions, people face a range of psychological risks, including burnout and compassion fatigue—also referred to as secondary traumatic stress or vicarious traumatization. These conditions lead to numbness, withdrawal, anxiety, difficulty sleeping, nightmares, anger, and cynicism. There is an emotional cost to caring genuinely for others. But there is also an inexplicable emotional gift. Caring for other people and animals can make us more hopeful and courageous—a notion called "vicarious resilience" in the field of psychology. Through the eyes of survivors, we can glimpse the promise of a better world. The inspiring stories of survivors—Phoenixes like Doc and Grace—provide insight into a better world that many people envision, have faith in, and work toward every day.

Survivors' stories have led me to extraordinary places, literal and figurative sanctuaries I call "Phoenix Zones." In Phoenix Zones across the globe, where the injured heal and mend, I've found remarkable similarities, not just in people's stories, but also in the tales of animals. In each sanctuary, I've found core principles—specifically, respect for basic liberties and sovereignty, a commitment to love and tolerance, the promotion of justice and opportunity, and a belief that each human and nonhuman animal possesses dignity. In Phoenix Zones, I've witnessed how practicing these principles can encourage resilience and even broader social change. Those who create these sanctuaries prove that it is within our power to change life for the better, and they give us the courage to do the same.

But first we must understand the roots of subjugation, violence, and suffering, the basis for our vulnerability and resilience, and all that Phoenix Zones are up against.

2 : UNEARTHING THE SHARED ROOTS OF VIOLENCE AND VULNERABILITY

You cannot cheat the law of the conservation of violence: all violence is paid for . . .
— PIERRE BOURDIEU, *Acts of Resistance: Against the Tyranny of the Market*

We live in a strange time, an era marked by contradictions. In recent decades, tremendous strides have been made toward social justice. Generally, our society has become less and less tolerant of prejudice and violence based on differences in race, gender, orientation, ability, and even species. People have become more aware of how others suffer, and many organized efforts address some of the most egregious forms of suffering in the world. But turn on any television, computer, or handheld device and you will still be inundated with tragic stories about the most vulnerable among us. The world seems to be unraveling, with reports of refugees struggling to reach safe ground—torture survivors like Doc and rape victims like Grace caught in conflict zones. Human trafficking and human slavery have once again become modern problems, as they were for Aiyana, while terrorism and other extreme forms of violence are increasingly perpetuated around the world. These human stories are compounded by other types of violent accounts—animals under siege in their own homes in the wild, neglected and abused like Love, and forgotten and unseen in industrialized farms and laboratories where they are subjected to unthinkable cruelty.

Why are so many still living in agony when awareness about their suffering is growing? What has given rise to these contradictions? Is it our reluctance to examine the roots of violence? Have we ignored the principal connections among different forms of violence and suffering?

If we look closely, there are pathological themes to much of the violence that plagues society. Domination, exploitation, and abuses of power are the causes of—the pathological explanation for—much of the malice around the world. Violence extends beyond physical abuse to include mental abuse, and beyond individuals to include institutions. Oppression and discrimination—tools of violence—deprive individuals of fundamental needs like liberty and justice, which can lead to physical and mental suffering.

These interconnected themes are not new. For centuries to millennia, marginalized mortals have been captured, abducted, enslaved, exploited, and abused simply because of their color, shape, size, or bend. Their bodies and minds have been unjustly robbed of independence and promise. Our historical record of violence is now on repeat, tripping over itself, playing the same phrase over and over again—birthed and bolstered by financial, political, and legal traditions. The pathological effects of violence are long and deep, leaving scars on our individual forms and on society. But, as with other types of pathology, we need to understand the entirety of the disease of violence—its roots and branches—to prevent and eradicate it.

ROOTS AND BRANCHES OF VIOLENCE

There is a clear link between violence against people like Aiyana, Doc, and Grace and violence against animals like Love. At times throughout history, it has even been treated as one problem. In the nineteenth century, among religious and secular advocates in the United Kingdom and the United States, the inhumane treatment of people and animals violated central mores like freedom, kindness, and fairness.[1] As a result, causes to protect people and animals united over a common moral vision. Organizations slowly emerged to protect the most vulnerable members of society—particularly children and animals—focusing on compassion for animals as a first step toward civilization. In the post-Civil War period in the United States, educators, feminists, and child protectionists, among other reformers, aligned with animal protectionists. Advocates emphasized religious creeds like mercy, care, and protection. Gradually, with increased secularization in society, their emphasis shifted to the importance of justice and respect for the rights and dignity of all people and animals. Leaders like Henry Bergh, of the ASPCA, wisely realized that since cruelty against children and animals shared common origins, policy measures and sanctions to protect them required similar responses and solutions.[2] Fortunately, by the early part of the twentieth century, laws had been introduced to safeguard children, though not animals.

Although the relative absence of laws to protect animals has obvious implications for animals, it also has consequences for people. A history of animal abuse is one of the most significant risk factors of who will become an abuser of an intimate partner, child, or stranger. Animal abuse is frequently a red flag for current or future family violence. As just one example, more than half of abused women have companion animals, and in half of these cases, their abusers have also hurt or killed their animals.[3] But the link between violence against humans and animals isn't limited to physical violence. It also extends to sexual violence. In one Australian study, a history of animal abuse was a better predictor of sexual assault than were prior convictions for homicide, arson, or firearms offenses.[4] As researchers have noted, abusers sometimes also threaten, harm, or kill their children's animals in order to coerce children into sexual abuse or remaining silent about abuse.[5]

Perpetrators often exploit victims' compassion for animals to keep them from seeking help. Studies show that women frequently delay leaving an abusive situation to protect animals in their care.[6] When they do leave, they often move to another unsafe situation, like a car, until they can find a women's shelter willing to take in their animals too.

Battered women and children could be particularly empathic to animals because they know what it feels like to be disrespected, demeaned, and degraded. They recognize their own vulnerability in animals. At the same time, some children who have been physically or sexually abused may grow up to be abusers of other people or animals. These tendencies could be explained by what researchers have called the "cycle of violence," much like a heritable disease, or the "contagion of violence," similar to a communicable disease.

Something larger than the heritability or contagion of violence is at work, however. Structural violence—an unjust and exploitive political and economic organization of society—trickles down to the individual through what sociologists call "the law of the conservation of violence."[7] Large-scale social forces and institutions can cause illness and death by preventing individuals from meeting their needs. For example, institutional racism and sexism can lead to afflictions at the individual and community levels. Norms, economic decisions, and laws that allow for the creation of suffering in animals also represent a form of structural violence that can affect animals and humans. As one example, domestic violence and child abuse are endemic to some communities where slaughterhouses are located. An extensive eight-year analysis of more than five hundred counties in the United States showed that slaughterhouse employment independently increased arrests for violent crimes, including rape and other sexual offenses.[8] These connections could be caused by the psychological trauma and desensitization associated with daily exposure to large-scale violence and death, but there could be other explanations.

In an interview for the *New York Times*, George Yancy, author of *Black Bodies, White Gazes*, spoke with Peter Singer, author of *Animal Liberation*, about the connections between racism, sexism, and biases against animals based merely on species identification.[9] Singer pointed out that it is difficult to reject one form of prejudice and oppression while accepting and practicing another. Any form of discrimination, tyranny, or abuse fosters an illusion that a free, sovereign, and just social order can be broken. In studies of children and adults, psychologists Kimberly Costello and Gordon Hodson have exposed how our treatment of animals directly influences how we treat each other.[10] For example, they've shown how widening the moral gap between people and animals—and treating animals as inherently lesser beings—can lead to racial prejudice.

A culture of disrespect lies beneath infectious constructs like misogyny, racism, child abuse, and animal abuse. Whereas the word "respect" comes from the Latin word "respectus," which means to look, and to look again, "disrespect" implies a failure to look at, or to consider, another. Undercover video has repeatedly caught farmworkers beating and anally penetrating pigs and other animals with rods while also denigrating women, girls, and people of color. Women's bodies have been referred to as "meat," much like the flesh of animals, or "bitch," like a female dog.[11] People of color have been called chimpanzees or monkeys in attempts to debase them.[12] Women, people of color, children, and animals simply aren't *seen* in these situations.

All of these links underscore the importance of addressing the deepest origins of discrimination and abuse—including violence directed at animals. If we neglect any branch of violence, it can take root like a metastatic form of cancer, spreading across the invisible lines we've created and placing us all at risk.

A COMMON VULNERABILITY

As many survivors realize, what links us all, even more than the pathology of violence, is our vulnerability. False forms of power characterized by domination can exploit it. Alternatively, vulnerability can be embraced and nurtured to foster true strength and power.

Vulnerability—from the Latin word "vulnerābilis"—reflects our susceptibility to being physically and emotionally wounded, whether by violence or other means. One of the earliest uses of the word was to describe the hero Achilles's heel in Homer's *Iliad,* a Greek epic poem depicting the events of the Trojan War. Legend has it that at his birth, Achilles's mother immersed him in the River Styx to make him invulnerable. In doing so, she held him by one heel that never touched the water. Later in his life, as a warrior in the Trojan War, Achilles remained unharmed by his enemies until Paris, son of the king and queen of Troy, shot an arrow into his heel. Achilles was mortally wounded, but only after Paris learned the secret of his heel from the god Apollo.

The legend of Achilles reveals how even the strongest among us are vulnerable. The portrayal of Achilles as a hero, juxtaposed with a belief that he embodied the grief of his people, is a reminder that strength and vulnerability live together in all of us. Vulnerability is our most common denominator. And, like us, animals are categorically vulnerable beings. In fact, much of our own vulnerability stems from the fact that we *are* animals. Like Love's vulnerability, animals' vulnerabilities reflect some of our core needs as social and biological creatures.

OUR SHARED VULNERABILITY TO PAIN

Beginning in the 1600s, the French philosopher and scientist René Descartes popularized the idea that animals were mere machines—free of thought, language, self-consciousness, or significant feeling. Their pain, even if present, was morally irrelevant to Descartes. In order to prove his point, he nailed live dogs' and rabbits' paws onto boards and cut the animals open. With their beating hearts exposed, he defiled and removed their organs.[13] He dismissed their whimpers as no different from deranged springs in need of oil.

Though Descartes ventured to use scientific experiments to prove his theories about how animals don't experience pain, science and common sense have since proven him wrong. We, as humans, feel pain because we *are* animals—and because it can keep us alive. Pain helps us avoid harmful situations and further tissue damage, since painful experiences usually trigger an immediate withdrawal or escape response. Painful experiences also help us prevent repeated injuries, if we learn to avoid sources of pain and are free to do so. Pain promotes the healing of injuries through various chemical mediators. But those in power sometimes exploit our capacity for pain to achieve other goals, as with torture or abuse.

The essential mechanisms that make it possible to respond to pain are present throughout the animal kingdom. Although we've known for decades that birds and mammals experience pain, more recent research has centered on how fish and invertebrates, like octopuses, perceive pain. Fish possess the same types of sensory receptors as birds and mammals, and the nervous systems and brains of fish are sufficiently complex for fish to experience fear, pain, and suffering.[14] Even invertebrates like octopuses have well-organized nervous systems that include brain centers focused on sensory analysis, memory, learning, and decision-making. These areas of the octopus brain have been compared with the cerebral cortices of vertebrates (our brains' decision centers).[15] That the capacity for pain is shared so widely across species proves how critical it is to our survival.

Scientists have inflicted many different kinds of painful conditions on animals, and these experiments have proven that animals, like people, experience acute or immediate pain, as well as a slow crescendo pain. Similar to people, animals inflicted with diseases can become severely fatigued and depressed. Like us, they also become anorexic, sleep deprived, and more sensitive to pain. This type of pain and discomfort, also called "sickness behavior,"[16] explains why we feel miserable when we have a cold or the flu.

People and animals also express pain similarly—through avoidance, crying out, aggression, and physiological responses such as accelerations in heart rate and increases in blood pressure. However, some animals vulnerable to predation may hide signs of pain to improve their chances of survival. Animals commonly hide their pain when people are present, much like children mask their susceptibility to fear and pain around some adults. In light of all the evidence we have about how children and animals suffer, they could be enduring even more than we realize. The animals Descartes hurt likely suffered even more than their whimpers indicated—not only physically but also psychologically, as Charles Darwin soon showed.

OUR SHARED VULNERABILITY TO PSYCHOLOGICAL DISORDERS

Within a couple of centuries, Charles Darwin had put many of Descartes's ideas about animals to rest. Darwin revealed how animals exhibit powers of deliberation, imagination, intelligence, empathy, love, and many other mental qualities that humans possess. Nonetheless, Darwin's views did little to reconcile how animals should be considered morally in society, including in decisions about their use in experiments. In fact, over the past several decades—drawing on Darwinian continuity theory—scientists have used animals to study pain, medical illnesses, mood and anxiety disorders, posttraumatic disorders, and other physical and mental conditions, neglecting the ethical implications of such complex capacities.[17]

Within the field of psychology, in the late nineteenth century, Russian physiologist Ivan Pavlov was one of the first scientists to experiment on animals. Pavlov is most known for his classical conditioning experiments, showing how dogs salivated at the sound of a bell they learned to associate with food. He is perhaps less known for using electric shocks to induce acute neurosis in dogs and other animals.

From the early to mid-1900s, other scientists drew on Pavlov's classical conditioning experiments to study mental disorders in more animals. Like Pavlov, they aimed to extrapolate their findings to humans.

Many of these experiments used electric shocks and food deprivation to induce neuroses, depression, and anxiety in cats, dogs, and chicks. Scientists explained how the animals crouched, trembled, howled, and cried. The animals even starved themselves in an attempt to avoid electric shocks and objects associated with the shocks.

By the middle of the twentieth century, researchers had begun studying attachment disorders in animals. In a series of infamous experiments in the 1950s and 1960s, experimental psychologist Harry Harlow used monkeys to show the importance of early attachment between a mother and child.[18] In Harlow's experiments, monkeys were socially isolated from their mothers and other monkeys and forced to choose a "surrogate mother" made of terry cloth or wire, which also provided access to milk. Unsurprising to anyone who has ever held a baby, the babies chose to spend significantly more time with the terry-cloth surrogate mothers than with the wire surrogates—even if they could get more nutrition through the wire surrogates. Later, Harlow demonstrated that the babies would also turn to the terry-cloth surrogate mothers for comfort and security when he or his research assistants intentionally terrorized them. When the terry-cloth surrogate mothers were removed from the monkeys' cages, the babies became rigid, rocked, and screamed—much like they would when taken away from their real mothers. Over time, the monkeys became severely disturbed and couldn't develop normal relationships with other monkeys. It became clear that the monkeys required more than sustenance. They also needed a safe environment and the love of their mothers and friends.

By the 1970s, Harlow had begun placing monkeys in what he called the "pit of despair," a stainless-steel isolation chamber, in order to induce depression. The monkeys became withdrawn and suffered from anorexia and emotional disturbances. When released from the pit, they had little to no interest in forming normal relationships with other monkeys, and when they wouldn't have sexual relationships with other monkeys, Harlow devised what he called a "rape rack" to force sex upon them. They became even more disturbed.

Many of these early psychology experiments, combined with a modern-day theory that mental illness is mediated by neurotransmitters like serotonin and dopamine, have driven contemporary psychiatric experiments. Today, animals are used in all sorts of experiments related to mood and anxiety disorders, compulsive disorders, and posttraumatic disorders, in addition to other forms of mental illness. Variants of PTSD, depression, and anxiety have now been described in monkeys, dogs, cats, mice, rats, and many other animals.[19]

Scientists still use electric shocks, maternal deprivation, and food and water deprivation to study depression and anxiety in animals. One of the most common experiments is the forced swim test, which is used to evoke depression symptoms like the loss of the ability to experience pleasure, and behavioral despair, also called learned helplessness. In the forced swim test, small animals are placed in water containers, with no opportunity for escape or rest. Some animals struggle throughout the entire session, whereas others despair, stop struggling, become passive, and float, moving only enough to keep their eyes and noses above water so they can keep from drowning. The animals also lose weight, like some depressed people do, and the architecture of their brain changes, making it more difficult to cope clinically and socially. The learned helplessness seen in these experiments is considered analogous to how some people who have been repeatedly abused also behave.

Most animal experiments used to study PTSD also use techniques introduced in early psychology experiments—from electric shocks to maternal deprivation to social isolation. But these experiments also incorporate newer methods, such as scaring animals with a direct attack by a predator, near-drowning, or bullying. After they are traumatized, the animals display signs of PTSD, including freezing, an exaggerated startle response, and memory deficits.

Even though years of research have shown how animals can experience pain, distress, and psychological disorders, they are still systematically used as experimental research subjects, for food, for clothing, and as entertainment. To the humans who use them, they are a means to an end. But to the animals who experience suffering, it is surely a form of torture. Through medical, psychiatric, and other forms of research, we know that people and animals share a capacity for pain, acute and chronic illness, and mental disorders. However—despite all we've learned about animals in recent centuries—some still draw a line between how we treat people and animals, assuming animals don't suffer the way we do, or that their suffering doesn't matter. But science suggests otherwise.

WHO SUFFERS?

Animals, like people, have subjective inner lives. They are thoughtful, emotive, and mindful. Though levels and types of suffering are inherently incommensurable, people and animals experience positive and negative emotions in analogous ways. Fear, sadness, anger, and joy enhance survival—for people and animals. As a result, we see attachment disorders, depression, and posttraumatic and complex anxiety disorders in a large number of diverse species deprived of their needs. Interference with personal and social needs often leads to physical and mental suffering, explained by changes in brain construction and function, and expressed through specific behaviors that are comparable, though not necessarily the same, across species.

Most of us have reliable intuitions about how others feel—known as empathy. Although it is impossible to know exactly how others suffer or who suffers most, some scientists and philosophers have suggested that children, animals, and adults without the capacity to fully understand what is happening to them could suffer more than other humans do.[20] Some people and animals may be more vulnerable to suffering because of their inability to make sense of their plight, escape, or alter their circumstances.[21] Brain size might also affect the capacity for suffering, though perhaps not in the ways some would assume. Having a smaller brain with fewer neurological structures could actually contribute to suffering, since sensitive beings with less organized brains may have less flexibility and more limited coping mechanisms.

Though we may never fully understand the minds of individuals other than ourselves, especially nonhuman minds, we can apply what we know. We are all inherently vulnerable to suffering. However, our minds and bodies do not alone determine our vulnerability. As the stories of Achilles, Grace, Doc, and Love show, our innate vulnerabilities can be manipulated and deepened, or embraced and lifted. Social, political, legal, and cultural factors can leave us more or less vulnerable, stronger or weaker, resigned or resilient.

PATHOGENIC VULNERABILITY

Feminist philosophers Catriona Mackenzie, Wendy Rogers, and Susan Dodds have described what they call a "pathogenic" form of vulnerability influenced by domination and political violence.[22] Pathogenic vulnerability is what Grace lived through in her homeland, where she was targeted since she was a woman, and what Doc lived through because of his political beliefs.

In medicine, the word "pathogenic" foreshadows the spread of disease if its core causes are not eliminated. As with the disease of violence, pathogenic forms of vulnerability can be eliminated only if we eradicate their roots. Otherwise, we all remain vulnerable to the communicable nature of exploitation and abuses of power. But we must also look to our treatment of animals as a marker of progress against the longest roots of pathogenic vulnerability. We won't fully dismantle the injustices humans suffer without deconstructing the same problems that lead to animal suffering.

Animals are susceptible to pathogenic vulnerability in many of the same ways humans are.[23] Even when animals are capable of expressing their preferences in ways we understand, they're subject to our whims, much like children still are in many areas of the world. Many of our legal, economic, and cultural paradigms render human *and* animal beings even more vulnerable than they already are. They are structural forms of violence that exacerbate vulnerability. For example, Aiyana and Love had no political power. Legally, Love was considered property and could be bought and sold—no different than the cars and trucks that passed him by on the road to Nairobi. Although Aiyana had legal rights, her father did not recognize them, and she was still bought and sold. Even when Aiyana and Love escaped their circumstances, they were completely dependent upon others—whether the people they encountered decided to be kind or malicious, attentive or neglectful. Like children, animals cannot always protect themselves, particularly if they are dependent on humans for their needs. Both Aiyana's and Love's innate vulnerabilities—their physical and emotional needs—were amplified rather than addressed with the sensitivity they required. They weren't protected from the neglect or abuse to which they were so particularly disposed.

Though it may be difficult to accept, in some ways animals are even more vulnerable than children, since most animals don't have advocates in positions of power. Many parents try to represent their children's best interests, often with their children's input. Laws require it, and all of us have been children. But it's virtually impossible for us to represent the best interests of nonhuman animals as well as we represent the best interests of another human being—simply because we cannot understand nonhuman minds as well as we can understand other human minds.[24] Animals are often made more vulnerable and subjected to suffering simply because we fail to take them seriously enough to try to understand them.

Fortunately, attitudes are changing. Across the globe, from the United States to China, more and more people believe that animals fit within a framework of social justice. A 2015 Gallup poll even showed that at least one-third of Americans believe that "animals deserve the exact same rights as people to be free from harm and exploitation."[25] Take a moment to think about the gravity of that assertion—and its potential implications. We're already on a more compassionate, empathic path than the one Descartes paved. But we need to think more broadly—not just about preventing suffering, but also about what people, animals, and society need in order to flourish.

TOWARD A MORE HOLISTIC VIEW

Around the time of the Second World War, clinical psychologist Abraham Maslow described the essential elements we need to thrive as individuals and as a society. Maslow was the son of Jewish immigrants who fled religious persecution in Russia in the early twentieth century. As a child, he was bullied because of his religious and ethnic background. He felt isolated from his peers, and his home life was stained with sadness and loneliness. As an adult, he struggled to identify factors that lead to well-being and resilience. Later in life he explained, "I was awfully curious to find out why I didn't go insane."[26]

Maslow studied human and animal behavior. He thought animals represented an honest view of what was fundamentally human—"the roots of human nature laid bare."[27] Early in his career, he worked with Harry Harlow experimenting on nonhuman primates, but he later took a very different approach. Unlike most psychologists at the time, Maslow studied renowned visionaries like Frederick Douglass, Albert Einstein, Eleanor Roosevelt, and Albert Schweitzer. He also saw value in studying positive emotions in animals—like the affectionate devotion of dogs and the lightheartedness of monkeys—rather than how they suffer when deprived of their most essential needs. Maslow concluded that, in order to be mentally well, we must fulfill a "hierarchy of needs," a pyramid of five hierarchically arranged goals. At the time, he believed only a few of these needs applied to animals. Like many scientists of his time, he was unaware of the breadth and depth of animals' capacities.

The most fundamental and basic four layers of Maslow's pyramid contain what he deemed "deficiency needs," including physiological needs, safety and security, friendship and love, and respect and esteem.[28] When these needs aren't met, we suffer physically and mentally. Physiological needs are quite basic: Our bodies crave air, water, food, sleep, sex, movement, and a normal temperature. Otherwise, we suffer from suffocation, thirst, hunger, and pain or discomfort. Safety, love, and esteem are psychological needs. When our safety is threatened, we can develop anxiety or posttraumatic disorders. If the need for love is threatened, we may develop attachment disorders or become depressed and anxious. Disrespect for our dignity and choices can lead to depressive symptoms, including learned helplessness.

In addition to deficiency needs, Maslow judged that in order to reach our full potential, we need intellectual challenges, joy, spontaneity, creativity, and the chance to contribute to the greater good. Basically, we need opportunity, hope, and promise. Since he first proposed his theory, countless empirical studies have endorsed the motivational power of these needs. However, studies have largely failed to confirm that one group of needs is more important than the others.

Intriguingly, many of the key principles I've found in Phoenix Zones echo Maslow's hierarchy of needs, as well as the moral vision held by nineteenth-century advocates working on behalf of children and animals.

Maslow appeared to be on the cusp of understanding that the needs he described are important to both people and animals. He observed that "truth, goodness, beauty, [and] justice" would ultimately be explained through biochemical, neurological, and hormonal mechanisms[29]—fundamental building blocks that we share with each other regardless of our differences. In the more than half a century since Maslow first published his theory, we now have abundant evidence that the vulnerability and resilience of people and animals are bound by fundamental needs. All of these needs are central to our well-being—and to social progress, which at least two of Maslow's study subjects understood. Albert Schweitzer and Albert Einstein espoused the idea that true social progress and justice would occur only when we end violence against people *and* animals. Albert Einstein once remarked that separating ourselves from the rest of the universe is an "optical delusion" imprisoning us—one that we can only free ourselves from by "widening our circles of compassion to embrace all living creatures."[30]

RISING FROM VULNERABILITY IN PHOENIX ZONES

Brené Brown is a researcher-storyteller who studies vulnerability. She speaks of its power, drawing from thousands of stories she's collected over time or, as she calls them, "data with a soul." Brown shows how the courage to be vulnerable can be the birthplace of strength and resilience.[31]

Over time, I've collected another dataset on vulnerability and resilience—from human survivors I've met and cared for over the years, the animals I've come to know, and the Phoenix Zones they've risen in. Through survivors, I've learned where strength is born and resilience lives. We're all vulnerable beings. We fear. We feel pain. We break. And we suffer. We can never know or do enough to be perfectly safe. Though we rarely take the time to acknowledge it, we live in a fragile world. Merely being alive leaves us with wounds and scars. But we are also strong.

Resilience isn't solely or forever determined by our genetics, our childhoods, or the best or worst of our lives. It isn't fixed. Resilience, like vulnerability, is a biological phenomenon influenced by the life, laws, and love in us and around us. This is the basis for the Phoenix Effect, which hinges on whether our vulnerabilities are nourished, as in Phoenix Zones, or exploited. The same is true for animals. As with vulnerability, much of our capacity for resilience stems from the fact that we *are* animals.

Our brains are phenomenally plastic. We learn. We develop. We regress. We grow. Although form and function are guided by genetic factors, our experiences help shape the way we adapt and cope. Just as our minds react to negative experiences, we're also shaped by positive experiences. If there is a biological foundation for psychological suffering, there must be a biological basis for recovery—similar to how physical wounds can heal after a painful physical injury. Though we're susceptible to physical and psychological suffering, we're also resilient beings searching for opportunities to build strength.

Take a moment to imagine a global society focused on well-being and resilience rather than violence—on creating sanctuary, particularly for the most vulnerable in society. We still have a long way to go before the world becomes a sanctuary for all, but in the meantime we are left with a central question: How can we each cultivate resilience in our own communities—even within a tumultuous, uncertain world?

PART TWO

The Quest

PHOENIX ZONES—PRINCIPLES IN ACTION

3 : LIBERTY

REFUGE FOR ASYLUM SEEKERS AND CHIMPANZEES

Those who deny Freedom to others deserve it not for themselves . . .

— ABRAHAM LINCOLN

While they were held captive, Doc and Grace lived in constant fear, under a seemingly endless threat that their bodies would be violated, their lives taken. Though the pain inflicted on their bodies was intolerable, the suppression of their basic liberties was equally damaging. Domination and abuse of their bodies, through physical force, was used in an attempt to control and alter their minds. Their yearning for an orderly and just world was hijacked by an abrogation of their freedom.

When torture victims are imprisoned, they are often deprived of sleep, sensory stimulation, water, food, and movement—the most basic needs described by Abraham Maslow. Doc and Grace were left thirsty and hungry, injected with pain and disease, confined in small, dark spaces, and subjected to incessant desolation.

Despite the physical toll of confinement, deprivation, and abuse, the mind often suffers most. Over fifteen years of caring for torture survivors, I've learned that the most enduring wounds are psychological. Lacerations, abrasions, and bruises fade with time, but scars etched on the mind are far more difficult to erase. Prolonged fears and sadness that take the form of mental disorders are commonly diagnosed after acute or repeated threats to our most essential freedoms. In order to survive, our brains morph, sometimes into forms that betray us through changes in our thoughts and behaviors.

MIND DISORDERS IN OUR NEXT OF KIN

Though capture almost broke them, freedom opened a window for Doc and Grace to escape the terror they lived through. Once the body is unleashed, the mind can slowly follow. Freedom offers a chance for rebirth for many survivors—for the Phoenixes they rise to become. It allows space to reshape their lives by weakening the invisible tether to a painful past. Through my work with animals, I began to gradually understand the biological basis for this change. I realized how basic liberties are foundational to the Phoenix Effect and creating Phoenix Zones where survivors can thrive.

One cold evening in 2007, I had an epiphany. Though I knew something about how animals suffer, it took more time to understand the depths of their psychological torment, their potential for resilience, and how their needs mirror our own.

That evening, I had asked a colleague to speak to some health professionals we were training to forensically examine torture survivors. My colleague was a psychiatrist who had spent many years with trauma survivors. Like me, she provided assessments of torture survivors seeking asylum in the United States. During the forensic training, we were cramped in a dusky room in a medical school building in the middle of winter. Most of us had just finished a long day with patients, but everyone was eager to learn. My colleague spoke softly to the group of doctors, psychologists, and students. I leaned in and listened closely as she described the neurological basis for psychiatric disorders and resilience. She talked in detail about the anatomical and physiological changes that occur in people affected by PTSD, depression, and anxiety and compulsive disorders.

As I listened to her, I began to connect the dots—the anatomical structures and physiological mechanisms she described are those found in animals. This basic likeness is what researchers point to in their controversial psychiatric experiments on animals—from Pavlov's experiments with dogs and Harlow's experiments with monkeys to more recent experiments with other animals. As she explained, severe stress can sometimes overwhelm normal protective physiological responses, which can cause persistent functional and structural changes in our brains—just as we see in animals. These changes can lead to avoidance or hypervigilance. Trauma can be re-experienced over and over again in the form of nightmares or flashbacks—or through mood and cognitive changes, leading to withdrawal or an absent, hollow mind.

As my colleague stood and talked at the front of the room, I developed an even greater awareness of how the animals in Pavlov's and Harlow's experiments suffered. As she explained, fear and anxiety serve as a first line of defense for people and animals. Our fearful responses depend on the activation of a small subcortical circuit, tucked below the main decision centers of our brains. Brain chemicals released in an emotional rush cause a traumatic experience to take root subconsciously. The amygdala, the brain's almond-shaped danger detector, absorbs traumatic details so a new threat later will subconsciously trigger an alarm. Like the immune system, which develops antibodies to protect us from infectious diseases, the brain's emotional hub remembers risks and stays on guard. Though this response is meant to be protective, to enhance survival, it can become disordered—as we see with PTSD or generalized anxiety disorder. The amygdala also works in concert with other small subcortical structures. For example, the hippocampus, a small horseshoe-shaped brain structure found in all vertebrates, is thought to be an epicenter of emotion, memory, and the fight-or-flight response. It could also help explain some of the similarities in mind disorders across species. Chronically stressed, traumatized people and animals have decreased hippocampal size—perhaps because of recurrently and chronically elevated levels of cortisol, the key hormone involved in our response to stress. Abnormal levels of cortisol can be toxic to certain areas of the brain and other parts of the body, like the heart. As a result, internment and deprivation can alter brain size, brain cell density, and brain structures involved in learning and memory, like the hippocampus. These mechanisms also help explain the association between chronic stress and heart disease. Such transformations occur in the minds and bodies of mammals like us, birds, and other animals.[1]

The day after the forensic training with my psychiatrist colleague, I had an idea. At the time, I was leading an effort to highlight the harms of animal research and how national and international polices should change as a result. Since animals are denied basic freedoms through their use in research and entertainment and other areas of society, I thought it was imperative to understand the impact that loss of liberty has on them, without hurting them further. Though the relationships between captivity and negative physical, social, and psychological effects have been the basis for many attempts to develop "animal models" of human mental disorders, few researchers have placed animals' severe abnormal behaviors in a clinical, psychiatric, or ethical context.

To begin, I specifically wanted to understand whether human diagnostic criteria for psychiatric disorders like PTSD and depression could be applied to chimpanzees, our closest living relatives. Like other people, I had a basic understanding of chimpanzees. I knew of their incredible intelligence and their talent for developing a culture they could pass on to subsequent generations. Their tremendous aptitude for strong family bonds, love, and aggression was detailed in the books on my shelves. Like us, they prefer to govern their own lives. If they are allowed to, chimpanzees make clear and deliberate choices. But even knowing all of this, I hadn't yet appreciated the fullness of their lives, or what they were capable of missing. So, with a group that consisted of primatologists, psychiatrists, clinical psychologists, and animal behavior experts, I embarked on a research adventure that took us across the globe.

At the time of our research, it was already well established that captivity, early maternal separation, social isolation, sensory deprivation, and physical injury were damaging to nonhuman primates, including chimpanzees. Psychiatrists and psychologists like Martin Brüne and Gay Bradshaw had already laid the theoretical and clinical framework for how disorders like PTSD might present in chimpanzees and other great apes.[2] Despite the challenges of making diagnoses of mental disorders in animals with abnormal early social lives, they showed how these disorders could mirror mental illnesses seen in humans. However, despite the many similarities between humans and nonhuman primates, it was still unusual to study mental disorders in animals using the terms and tools of human psychiatry.

Over the course of two years, we took a method developed by child psychiatrist Michael Scheeringa that had been used to diagnose PTSD in young, nonverbal children and adapted it for use with chimpanzees.[3] To develop better criteria for the assessment of PTSD in children who couldn't report their own symptoms, Scheeringa relied on behavioral observations and reports from people who knew the children best. In human and veterinary medicine, doctors commonly rely on accounts from parents and guardians about young children, adults with cognitive impairments, and companion animals. Rather than causing more harm by intentionally eliciting symptoms of trauma, as some researchers do, we wanted to use harmless, noninvasive methods like those Scheeringa used.

Like the children in Scheeringa's studies, the chimpanzees in our study could not verbally communicate with us. However, as with children, we were aware of reliable, established behaviors classified as normal or abnormal in chimpanzees. We also knew that chimpanzees, like humans, are self-aware and understand cause and effect. Young chimpanzees inspect and manipulate objects that don't work as they'd expect, and they can improvise solutions to complicated problems. Learning and memory develop rapidly during the first few years of life, and young chimpanzees can even outperform adult humans on complex short-term memory tests. They also have long memories. Even after decades have passed, chimpanzees can remember specific places and objects. They remember people years after they've met them. For example, in his book *Next of Kin,* primatologist Roger Fouts recounts a story of Booee, a chimpanzee he taught American Sign Language.[4] Booee had been raised by a human family before he was transferred to Fouts's care. Within a few years of being with Fouts, Booee was moved to a laboratory facility in New York, where he was confined for more than a decade and infected with hepatitis C. At the request of the television program *20/20,* Fouts visited him there. When Booee looked through the white bars of his small cell, he recognized Fouts and signed and played with him. Other visitors reported that Booee used the American Sign Language gesture for "keys," indicating that he wanted out.

Pulling together all that my team knew of chimpanzees, we understood they were theoretically capable of developing PTSD and other forms of mental illness. Soon thereafter, my chief colleague in the study, primatologist Debra Durham, and I traveled to sanctuaries around the world, collecting data on chimpanzees. From Florida to Washington State, and from Japan to Kenya to the Netherlands, we met chimpanzees who had been subjected to years in confinement. We also made our way to Uganda, where we ventured into the national forest to gather data on chimpanzees living in the wild. We worked with field trackers who spent every day in the forest. They could identify chimpanzees by their individual personalities and characteristics, including the freckles on their face. In total, we studied more than three hundred fifty chimpanzees.[5]

Through our research, we met chimpanzees whose stories were reminiscent of those of the human torture survivors I had met and cared for. They had been trapped, stolen from their families, chained and confined, and severely abused. They had been deprived of food, water, sleep, and movement—the basic freedoms Maslow outlined.

We found that the prevalence of psychiatric disorders in chimpanzees who had been used in biomedical research, the entertainment industry, or the international primate trade was extraordinarily high. The psychological disorders seen in these chimpanzees were rarely, if ever, seen in chimpanzees freely living in the forest. We found that forty-four percent of the ill-treated chimpanzees displayed signs of PTSD, compared with only one-half percent of chimpanzees living in their normal habitat. More than fifty percent of the confined, maltreated chimpanzees met the diagnostic criteria for depression, compared with three percent living in the wilderness. When we looked more closely at the few chimpanzees in the forest who showed signs of mental disorders, we learned that they had been trapped by poachers. Their freedom had been compromised. Later, we showed that the traumatized chimpanzees had higher rates of anxiety and compulsivity, compared with chimpanzees living freely.[6] Chimpanzees who had been kept in laboratories were more likely to withdraw socially and lose interest in playing or eating. Some became excessively hypervigilant or aggressive, and others displayed repetitive, self-destructive behaviors.

THE CLE ELUM SEVEN

Today, all of the captive chimpanzees in our study live in sanctuaries, where they will stay for the duration of their lives. One of the sanctuaries Debra and I visited for our study was Chimpanzee Sanctuary Northwest, located on twenty-six acres of land about ninety miles east of Seattle, in a small town called Cle Elum. There, Diana Goodrich and her husband, J. B. Mulcahy, a coauthor on our study, have assumed the lifetime care of seven chimpanzees. Before settling in Cle Elum, Diana pursued degrees in psychology and public policy, worked with children with special needs, and led research on chimpanzee communication. J. B. began working with chimpanzees in 1998. His initial fascination with ape language studies led him to work with Roger Fouts at the Chimpanzee and Human Communication Institute, where he and Diana met. After learning about the plight of chimpanzees, their focus quickly turned from research to care and advocacy.

In June 2008, J. B. and Diana met the Cle Elum Seven: Annie, Burrito, Foxie, Jamie, Jody, Missy, and Negra. Before their journey to the sanctuary, they were kept in a Pennsylvania laboratory facility with no access to the outdoors. They were all used in vaccine experiments. Before that, some of them were captured in Africa as infants, brought to the United States, and kept as pets and used in entertainment. All of their babies were taken from them shortly after birth, to be used in entertainment or experimentation.

During the summer of 2009, after we launched our international study, I met the Cle Elum Seven. I eagerly walked up a steep driveway toward the sanctuary, which is situated on a large hill overlooking the Cascade Mountains. Bing cherry trees adorned the front of the acreage, and we watched as horses grazed in the neighboring meadow peppered with freshly cut grass. As I first strolled up a rolling hill with J. B. and Diana, I passed a cage. I later learned it was the cage the chimpanzees were transported in—and the normal size of laboratory cages for chimpanzees. The difference between that laboratory cage and the living space they have now is extreme. The cage was barren. It was smaller than an elevator car and had vertical silver bars as walls.

At the time of our study, about a year after they had been in Diana and J. B.'s care, all seven chimpanzees showed signs of depression, PTSD, anxiety, or a compulsive disorder. When I first met them, I was struck by their appearance. When they arrived at the sanctuary, the chimpanzees had far less hair than they do now, perhaps because they had compulsively pulled their hair in the confinement of the dark laboratory basement. J. B. and Diana told me that just after they arrived at the sanctuary, the chimpanzees' faces all looked like ghosts. When they described the chimpanzees' faces, my mind immediately went to the classic picture of "shell shock" widely publicized after combat veterans returned to the United States from the Vietnam War.

Each of the seven chimpanzees has their own idiosyncrasies, just like human beings. Foxie carries around troll dolls, and some sanctuary supporters wonder if the troll dolls symbolize the babies who were taken from her. Chimpanzees develop very strong maternal bonds with their infants, and infants rely on their mothers for the first several years of their lives. Normal chimpanzee family dynamics and social networks play a critical role in infant survival, cooperative behavior, social learning, and forming cultural traditions. When they are young, chimpanzees even pretend to be parents by playing with sticks in the same way young children play with dolls.[7]

The oldest of the Cle Elum chimpanzees is Negra. Sometime around 1973, in Africa, when Negra was only an infant, she was taken from her own mother, who was likely killed when she was forced to surrender Negra. Chimpanzee mothers rarely give up their babies freely or without a fight. After Negra was captured, she was shipped as cargo to the United States, where she was forced into a life of imprisonment, exhibition, and experimentation. She was leased to White Sands Research Center in southern New Mexico in the spring of 1982, where she was tattooed with the letter-number combination "CA0041." Just one month after her arrival, Negra was forced to breed with a chimpanzee named Mack. Other times, she was kept in isolation or with one other female chimpanzee. Negra's life in a laboratory was very different from the way her mother and father had lived in the African bush, where chimpanzees live in groups of families and friends, often separating and coming back together again as they wish, not unlike human societies.

Over the course of her life, Negra was continually injected with diseases and forced to undergo repeated biopsies. She was also impregnated. When she gave birth, every one of her babies was taken from her and put into separate experiments. She gave birth to her first daughter, Heidi, in January 1984, and Heidi was taken away from Negra immediately. Negra was able to stay with her second daughter, Angel, for only five days before Angel was taken away.

In 1986, Negra was placed in solitary confinement after a clerical error showed she had abnormal laboratory tests and the laboratory personnel decided she should be isolated from other chimpanzees. Negra was kept in isolation for years. She was finally released from seclusion when the laboratory staff discovered her liver tests were completely normal. Soon after she was released, she was impregnated again and the third of her infants, Noah, was also taken away from her. Though I met Negra after she was released from her imprisonment, it quickly became clear that years of pain and isolation continued to haunt her.

Even before I met her, I knew Negra had suffered deeply over the course of her life. I read the few medical reports about her that were available. She felt the pain of a needle, the agony of inflicted diseases, and a psychological ache I was beginning to understand. When I met her in 2009, Negra was in her late thirties, about my age at the time.

In our study, Negra met the diagnostic criteria for PTSD and depression. As I watched her, she sat solemnly with a pink and blue blanket partially covering her head. She kept to herself in a nest of blankets separate from the rest of the group. She slept excessively during the daytime, and she was uninterested in playing, eating, grooming, or socializing with other chimpanzees. She was slow and sluggish. If someone touched her unexpectedly, she would scream or run away. And, though her caregivers and veterinarians searched and searched, there were no medical explanations for her abnormal behaviors.

A DOSE OF FREEDOM

Negra's forever home is now the sanctuary—hundreds of miles away from that small laboratory cage, what was once called "Cage #28." With time and distance from the confinement of the laboratory, over a period of several years, Negra's symptoms have subsided. Her behaviors have normalized. She has spent more time exploring outside. She loves peanuts and fresh spring grass, and she often greets her caregivers in the morning with gentle kisses on their wrists. Over time, she has gradually bonded with Burrito, started to play chase with Diana, wrestled with Missy, and reintegrated with the others. Gradually she has also welcomed the touch of her human caregivers. One of her caregivers was nearly in tears when Negra first allowed her to give her a knuckle rub—a signal of trust. But from time to time, Negra still endures restless sleep and what appear to be nightmares. After she accidentally bumped into an electric fence that shields the chimpanzees from the rest of the world, for a few moments Negra was filled with fear. And some sounds still trigger bad reactions from her—a reminder of how difficult it is to diminish the mind's scars.

Nonetheless, Negra doesn't shy away from standing up for herself, or asking for what she wants. As I learned when I first visited the sanctuary, Negra loves to watch people dance, and she and the others will tell you if they approve of your dancing by nodding their heads and signaling with their hands to go on. After my first dance—a combination of the Smurf, Electric Slide, and other moves—I was delighted to receive requests for an encore. I didn't dare stop, even when their caregivers looked at me like I was enjoying it a little too much. As one caregiver said about Negra, it's thrilling that she finally gets to speak her mind and be heard with love. Unlike what the Cle Elum Seven experienced in the laboratory, their caregivers are respectful of the ways they express themselves—a newfound freedom.

In the summer of 2016, I returned to visit the Cle Elum Seven with my friend John Gluck, a former primate researcher who worked with Harry Harlow years ago. Today, John is an animal protectionist, and he details his ethical journey in his memoir *Voracious Science and Vulnerable Animals.*[8] John and I listened as Diana and J. B. described how Negra had continued to improve. She is particularly fond of prickly lettuce, a weed that grows like wildfire beneath a shaded area at the top of what's now named Young's Hill. Until recently, Negra had been timid about venturing up the hill. At times, I have wondered if Negra truly feels the freedom she has, or if her mind is still working its way out of that basement she was kept in for years. But there are signs she sees her window of freedom. Now Negra climbs up Young's Hill, takes her time gathering prickly lettuce under a pile of old lumber, and brings it down the hill to enjoy among her friends.

With the help of volunteers, staff, and supporters, Diana and J. B. have given the chimpanzees a good life. They have given the chimpanzees liberties they were robbed of for decades, safety from the harm of experimentation and exhibition, love, and a respect they seldom before received. They consistently prioritize their needs, show them kindness and affection in ways that don't violate their independence, and play and groom with them. J. B. and the chimpanzees can often be found playing hide-and-seek, chasing each other around the sanctuary. Though a fence separates them from J. B., they easily find ways to express their affection for one another. The love J. B. and Diana have for the chimpanzees is echoed by the fondness the chimpanzees have for them. They are also now free to express their love for each other. I got goose bumps the first time I saw chimpanzees in our study hugging and comforting each other, after years of barely touching each other through the bars of small cages separating them in a laboratory.

Much of what the Cle Elum Seven need and receive at the sanctuary is supported by scientist Sue Savage-Rumbaugh's research. In conversations with another group of chimpanzees who can communicate through sound, motion, and symbols, Savage-Rumbaugh asked them about what they desire in life. She compiled at least a dozen needs—including self-determination, independence, the freedom to secure their own food, freedom from fear of human beings attacking them, and the freedom to maintain lifelong contact with those they love, make unique contributions to their group, and pursue new experiences. She then asked the chimpanzees if these needs were important to them. Using established methods of communication, they responded, yes, they are.[9]

After years of living in confinement, Negra and her friends are no longer living under constant fear or the threat of bodily harm. The changes in Negra's behaviors reflect what is happening in her brain. With time, even after years of trauma, our—and their—brains can rebound. With time, derangements in stress hormone regulation can normalize, even allowing brain structures to return to a more normal shape and size. For example, after release from captivity, the hippocampus—that small horseshoe-shaped structure in the subcortex of the brain—can expand again. This transformation reflects changes in brain cell density and an absence of the toxic effects of brain chemicals like cortisol released during unremitting stress. The brain's plasticity—responding to freedom—provides at least some of the basis for the Phoenix Effect.

Changes in neuroendocrine hormones, like serotonin and norepinephrine, which antidepressant drugs target, may also be involved. Elsewhere, some researchers and caregivers have tried to treat depression and PTSD in chimpanzees with selective serotonin reuptake inhibitors and serotonin norepinephrine reuptake inhibitors, medications used to treat the same disorders in humans.[10] Though these medications sometimes help—as they sometimes do in humans—they pale in comparison to the effects of recovered freedom.

THE LAST 1,000

Unfortunately, the Cle Elum Seven are not alone. For almost a hundred years, since the 1920s, chimpanzees have been used in the biomedical research industry in the United States, the last industrialized nation to experiment on our next of kin. In tribute to chimpanzees who remain in laboratories, feminist scholar Lori Gruen established a website called "The Last 1,000."[11] Many of the chimpanzees featured on her website were orphaned when their mothers were killed so they could be trafficked to the United States as infants. Once in the United States, many of these sensitive beings were used in space program rocket sled and decompression experiments and later subjected to water, food, sensory, and sleep deprivation, illicit drugs, toxins, and deadly viruses. On Lori's website, the chimpanzees' names turn green when they are released from laboratories to sanctuaries. Their names are listed in alphabetical order. Some are known only by a tattooed number or location—from Atlanta, Georgia, to Alamogordo, New Mexico, where a fight for the lives of all chimpanzees in US laboratories was launched.

In 2001, a group of almost two hundred chimpanzees who had survived space and laboratory experiments were placed in a colony in Alamogordo, New Mexico, where they lived in small groups and were kept out of experiments for a decade. In 2010, the National Institutes of Health (NIH) announced a decision to move the Alamogordo chimpanzees to a private San Antonio, Texas, research facility, where the chimpanzees—many of them elderly and sick—would be used again in invasive experiments. I've pored through many of those chimpanzees' medical records. They remind me of many of my own sick patients with heart disease, congestive heart failure, diabetes, and end-stage liver and kidney disease. Many years of injected diseases, being shuffled back and forth between public and private laboratories, and captivity have taken their toll.

The public responded to the NIH announcement with outrage. Opposition to moving the chimpanzees was especially strong since the announcement followed worldwide bans on chimpanzee experimentation by many governments and pharmaceutical companies. Public officials and scientists joined ranks with activists and a growing number of citizens who opposed chimpanzee experimentation. Following a request by Senators Jeff Bingaman and Tom Udall from New Mexico and Senator Tom Harkin from Iowa, and a similar request from former New Mexico governor Bill Richardson, the NIH commissioned a report by an Institute of Medicine committee to produce an analysis of the scientific necessity of chimpanzee research. In December 2011, the committee issued its final report.[12] Though the report did not endorse a ban on chimpanzee research, it established a set of criteria for determining when, if ever, current and future use of chimpanzees in research would be deemed "necessary." Their deliberations were not without drama. Several original members of the committee were removed due to perceived conflicts of interest. The committee was asked to ignore moral and ethical questions about the justification of chimpanzee research, but it decided that was an impossible request.[13] At times, the committee speculated about whether guidelines for chimpanzee research should resemble those for children or human prisoners, who receive enhanced research protections under US law.

In the end, the committee concluded that chimpanzees were largely unnecessary for research. Their recommendations were uncommonly demanding guidelines for federally funded research involving animals. Unlike virtually all previous top-level reports like it, NIH director Francis Collins accepted it as federal policy within two hours of the report's public release. Within two years, Collins announced that federally funded chimpanzee research would be phased out in the United States. At the time, the fate of many of the chimpanzees in laboratories was unclear. Would they be released to sanctuaries or remain in the laboratories where they had borne the pain and suffering of laboratory protocols?

Fortunately, there have been further positive changes for chimpanzees. Unlike before, chimpanzees living in both captivity and the wilderness are now listed as endangered under the Endangered Species Act. The Fish and Wildlife Service previously recognized chimpanzees living in the wild as endangered, but captive chimpanzees were deprived of such protection—so virtually anything could be done to them in a laboratory, zoo, or home. In effect, this policy change rigorously restricts the use of chimpanzees in laboratory experiments, interstate trade, and the entertainment industry.

Nonetheless, many chimpanzees are still waiting for their day of freedom. They are waiting for sanctuary. Opponents of their freedom have worried about a slippery slope—which many people frankly hope for. Just as there is a slippery slope from violence against animals to violence against people, there is also a slippery slope toward compassion for all beings.

Many people believe international and national laws already protect animals from severe pain and suffering. But this is not true. Chimpanzees can still be kept in an enclosure like Negra's "Cage #28." Some animals are not even considered "animals" under the only federal law in the United States that regulates the treatment of animals in research, exhibition, transport, and trade.[14] Many animals—like mice, rats, and birds used in laboratories and animals used in food production—are intentionally kept off the books, and almost anything can be done to them, contributing further to their vulnerability. For animals like Negra, covered under federal law, there is still no threshold for what can be done to her in the name of science. Today, Negra's largest source of protection comes from the fact that she is owned by people who care about her well-being, rather than what she can offer them in the name of science or entertainment.

THE LONG PATH TO FREEDOM

The minimal protections afforded animals differ substantially from laws that protect people. But human rights were not magically bestowed upon us. They were enacted only following fight after fight—and they are still under threat today. As Doc once reminded me over tea, though ideas about human rights have been around for centuries, freedom is never guaranteed. Our legal and political histories lay bare the truth of his remarks.

At the heart of every human rights resolution is a conviction that we should not be unjustly imprisoned or suffer torture and other trespasses—that we should be free from the most egregious abuses, including confinement. Legally, these rights are referred to as bodily liberty and bodily integrity. These rights were violated when Doc, Grace, and Aiyana were captured and abused. Most human rights resolutions cover far more than the freedom to move at will and avoid abuse, but they start with these fundamentals, almost in homage to our most basic needs.

Although Cyrus the Great of Persia is credited with introducing the first human rights charter over twenty-five hundred years ago, the second internationally recognized human rights charter was drafted after the French Revolution—though it excluded women and did not revoke slavery. Within two years of the birth of the United Nations, the third and most recent international human rights charter, the Universal Declaration of Human Rights, was drafted. Responding to atrocities witnessed in the Second World War, the Universal Declaration declared all humans free and equal in dignity and rights, transcending race, religion, sex, class, political opinion, and national or social origin. Unlike the French Declaration, the Universal Declaration on Human Rights prohibited slavery. Considered revolutionary at the time, the first four of its thirty articles actually mirrored the original human rights charter conscripted in the first millennium by Cyrus the Great. And though the charter prohibited torture, that battle had not yet been won. Perhaps it still hasn't.

Through my work with torture survivors, I've become engrossed in the legal and political history surrounding torture—one of the gravest violations of freedom.[15] In 1973, Amnesty International released its "Report on Torture," warning of a trend toward the systematic internationalization of torture, styling torture as a "social cancer."[16] Within two years, the United Nations had adopted the Declaration on the Protection of All Persons from Being Subjected to Torture and Other Cruel, Inhuman, or Degrading Treatment or Punishment. Nonetheless, torture continued across the globe. But in 1984, a second Amnesty report revealed a pattern of institutionalization of torture in militaries, once again rallying the international community against torture.[17] People responded in outrage to torture techniques with euphemistic descriptors like "parrot's perch" (a form of suspension), "the grill" (electric shock), and "the bath" (akin to waterboarding).

Finally, in 1984, by a unanimous vote, the United Nations approved the Convention against Torture. Torture was defined in detail, leaving little room for misinterpretation, as "any act by which severe pain or suffering, whether physical or mental," is intentionally inflicted to obtain information or a confession, punish, intimidate, or coerce, "based on discrimination of any kind . . . at the instigation of or with the consent or acquiescence of a public official or other person acting in an official capacity."[18] The Convention allowed for no exceptions, even during a state of war, threat of war, or public emergency. An order from a superior officer or public authority could not be invoked as justification for torture.

Despite historical advancements, there are still significant threats to freedom from torture and other trespasses—dangers to our bodily liberty and integrity. In the aftermath of the September 11, 2001, attacks, the Bush administration reversed decades of progress against torture. However, the Bush administration was not alone. Members of Congress, political pundits, and Ivy League professors mobilized to justify torture.[19] Some prisoners were interrogated twenty-four hours a day while they "cried, begged, pleaded, and whimpered," according to a 2014 US Senate Intelligence Committee report on the Central Intelligence Agency's (CIA) use of enhanced interrogation techniques.[20] Detainees were subjected to hypothermia, solitary confinement, sleep and sensory deprivation, and sexual assault. At least one man, Gul Rahman, died of complications of hypothermia at the notorious "black site" in Afghanistan known as the Salt Pit, after he was shackled and left half-naked in a cold cell. He had been arrested with a physician who was also left naked, restrained, and suspended in the Salt Pit.[21]

And in 2016—despite the 2014 report's conclusions that torture does not help secure intelligence—some presidential candidates heedlessly revived calls for torture. These calls for torture reflect what Alfred McCoy elaborates in his book *Torture and Impunity*: that there has been "thin discourse," "evasion, euphemism, and denial," around the problem of torture, thereby threatening our most basic liberties.[22] As they were in the 1980s, euphemisms used to describe torture have become the archenemy of moral clarity—designed to blunt our ability to view moral problems clearly, while suppressing our emotion and intelligence, and justice.

It's worth noting that euphemisms are also commonly used to describe how humans treat animals. For many years, euphemisms clouded discourse around the use of chimpanzees like Negra in laboratory experiments. Electric shocks were called "negative reinforcement" or "noxious stimuli," forced isolation was termed "single housing," and severe pain and distress were referred to as "stressful research protocols." But what would we call a nonconsensual electric shock to a human, other than torture? Taking baby chimpanzees away from their mothers, siblings, and peers was labeled "nursery rearing" and "limited socialization." In humans, these acts would be called kidnapping and solitary confinement. Still, softened phrases are used today when applied to other animals. In effect, euphemistic rhetorical devices have become ethical blinders to the complex suffering of people and animals. But when we push through these blinders, and the bias and prejudice that often lie beneath, we can make tremendous strides. We might even create a freer society characterized by compassion and justice—rather than violence and injustice.

WHAT THE FUTURE HOLDS

The struggle toward securing and enforcing the right of every human being to be free isn't over. However, international human rights laws at least reflect the right to liberty and other rights we need to be healthy and well. This isn't the case for animals, begging the question of whether animals like Negra should have a legal right to freedom as humans do.

A lawyer, and friend, by the name of Steven (Steve) Wise is fighting to end this discrepancy—by seeking two fundamental rights for his nonhuman plaintiffs: bodily liberty and bodily integrity, the same language used in human rights law.[23] For a chimpanzee, respect for bodily liberty would exclude spending life in a laboratory, and respect for bodily integrity would prohibit being forcibly inseminated or injected with a lethal disease. Steve makes the point that we should not confuse these fundamental rights with human rights: "Human rights are for humans. Chimpanzee rights are for chimpanzees. Dolphin rights are for dolphins. Elephant rights are for elephants."

In late 2013, Steve brought the first of several lawsuits to courts in New York, attempting to establish basic rights to freedom for four chimpanzee plaintiffs. One of his first plaintiffs was Tommy.

Before he brought the case before the New York courts, Steve met

Tommy on a compound near the Adirondacks.[24] On an early October evening, Tommy looked beyond the iron bars of his small cell to see Steve, the man representing him. Tommy kneeled in what Steve described as a dungeon. For years, Tommy lived in solitary confinement in a small, dark cage in a used-trailer lot—with only a television and cement walls painted with a jungle theme. Two months after his visit to see Tommy, Steve legally petitioned the court to release Tommy from solitary confinement. In court, he presented exhaustive scientific evidence from the world's leading primatologists about the capacities and needs of chimpanzees.

On December 4, 2014, the Supreme Court, Appellate Division, Third Judicial Department, issued its decision. Tommy would not be released from what Steve's legal team refers to as false imprisonment. However, the judge was not without sympathy. Fulton County Supreme Court Judge Joseph M. Sise concluded: "I am sorry I can't sign the order, but I hope you continue. As an animal lover, I appreciate your work."[25] Steve and his team appealed the decision and later addressed a judge in New York's highest court, who didn't decide in Tommy's favor but gave the legal team a few more steps to stand upon.

After the trial, Tommy's whereabouts became unknown. The people who own him hid him from authorities. As Steve points out, despite his status as a member of an endangered species under federal law, it is still nearly impossible to reliably track Tommy's whereabouts—that's the difference between being "something" and "someone." This gap is what the Nonhuman Rights Project is fighting to change. Meanwhile, Steve has continued on with cases on behalf of other chimpanzees—including Kiko, a chimpanzee owned by a couple in Niagara Falls.

If Steve is successful in the future, chimpanzees like Tommy will be released to sanctuaries in North America that reflect their needs. Having visited the sanctuaries where they could receive refuge, I can attest that the difference between their current cells and the freedom they would experience in a sanctuary is like night and day. In the sanctuaries, the animals are the point of focus. Their needs, not the whims of those caring for them, are what matter. If an appeals decision favors any of the chimpanzees Steve represents, it could set a precedent—once and for all—to provide freedom for chimpanzees like Tommy.

Steve and his legal team are also preparing to file suits on behalf of confined and abused elephants, dolphins, and whales. Like chimpanzees, these animals clearly demonstrate a capacity for agency: when individuals act independently and make their own choices. Using legal arguments as old as the ancient Romans and Greeks, Steve has made the case that the first legal cases on behalf of animals should be based on their capacity for autonomy—self-determination, self-agency, self-knowledge, and self-consciousness. All the while, he and his team have remained strategic and transparent. He has even written law articles and books outlining his legal strategy.[26] When he's not in court, he debates renowned attorneys around the globe to strengthen his legal arguments.

Though none of the lawsuits brought by Steve have yet resulted in freedom for Tommy or others, there is hope. In response to a case filed on behalf of Kiko, Niagara County Supreme Court Judge Ralph A. Boniello III ordered a telephone hearing and concluded, "I have to say your papers were excellent. However, I'm not prepared to take this leap of faith . . ." to declare that a chimpanzee should be granted the legal right of habeas corpus.[27] One day, all it will take is a leap of faith—as it has for many human beings who have been wrongly captured, abused, or tortured. It will require a leap like the one taken on behalf of the abused child Mary Ellen, whom Henry Bergh and his ASPCA colleagues represented in court. In another hearing, Justice Barbara Jaffe asked the question at the heart of each of these cases: "Isn't it incumbent on the judiciary to at least consider whether a class of beings may be granted a right or something short of the right under the habeas corpus law?"[28]

Steve's journey to Tommy and the other chimpanzees has been long and determined. He is a modest man, originally drawn to medicine and human rights because of his interest in social justice. Though he isn't sure if he will see the day when the law changes, he teaches his law students animal rights jurisprudence—a form of law that doesn't yet exist—with an expectation that they will one day live in a world in which a body of law protects the freedom of animals.

Steve has based his legal strategy on legal arguments that ended human slavery. For decades, he studied what it took for enslaved humans to secure basic rights to their freedom. In his book *Though the Heavens*

May Fall, he details his research on Granville Sharp, a civil servant and political reformer.[29] In 1769, Sharp printed *A Representation of the Injustice and Dangerous Tendency of Tolerating Slavery.*[30] At a time when many abolitionists argued that slavery was wrong because of the horrible conditions enslaved people were kept in, Sharp argued that the very nature of slavery was wrong, since it violated bodily liberty. In 1771, Sharp befriended James Somerset, an enslaved man who had escaped a ship used to traffic people from Europe to the British colonies in the Caribbean. He brought Somerset's case before the Lord Chief Justice of England, and Somerset eventually won his freedom. Many regard the decision as effectively abolishing slavery within Britain.

Had Granville Sharp been alive today, I believe it's possible he would have joined Steve's legal team. In his memoirs, Granville Sharp wrote about his affinity for animals and his aversion to the cruel ways humans treat them.[31]

ONE FREEDOM

One of my friends who restores natural habitat for chimpanzees in eastern and central Africa once asked me if we recognize civil rights for animals in the United States. Coming from a country rocked by political corruption and violence, his question initially surprised me. When we talked more, I began to understand why he asked the question the way he did. He assumed that since our nation legally recognizes basic human rights—regardless of race, gender, or political affiliation—we would also recognize the fundamental rights of at least some animals. For him, recognizing the rights of animals to basic liberties falls on a continuum with basic human rights.

At times, it's difficult to understand why we haven't made more rapid progress toward acknowledging the basic rights of marginalized people and animals around the world. Are we afraid? Does recognizing others' rights to basic liberties threaten our own? Is it possible that those who are most threatened by progress fear losing their counterfeited threads of power? Equality could feel like oppression to those with privilege.

But these fears are unfounded. Domination is a fabricated idea, one that can be easily pierced. When the freedom of a single being is compromised, we are all at risk. To the contrary, a universal commitment to freedom could better protect us all from bodily trespasses cast against us. Throughout history, we've already witnessed how expanding the fight for liberty from a few to many deepens and reinforces all of our freedoms.

4 : SOVEREIGNTY

GLOBAL SANCTUARY FOR UNCHAINED ELEPHANTS AND PEOPLE

Over himself, over his own body and mind, the individual is sovereign.
— JOHN STUART MILL, *On Liberty*

When I revisited the Cle Elum Seven in the summer of 2016, Diana shared a heartrending series of reflections. To meet legal requirements, many barriers, including bulletproof glass and an electric fence, separate the sanctuary's chimpanzees from the surrounding community. The chimpanzees can swing from a platform made of fallen timber, climb over a large reddish-brown rock, and sit atop Young's Hill with panoramic views of evergreens, a yellow-green meadow, and a rushing river. They can rest with their friends, or they can seek solitude. When they wish, they can retreat to an indoor enclosure. But the humans—not the chimpanzees—hold the keys to the locks that separate the Cle Elum Seven from the rest of the world. Though the chimpanzees have won a measure of freedom, they aren't completely sovereign. Their freedom is limited by factors outside their control.

Diana believes Jamie, one of the Cle Elum Seven who spent almost ten years in a trainer's home before being confined in a laboratory, is acutely aware of the limits to her freedom. Now considered the leader of the group, Jamie tries to exert as much control as she can over her own life and the lives of others. As one example, Jamie loves cowboy boots. People send her boots from all over the world. Several times per day, Jamie coaches Diana and other staff or volunteers to put on a pair of boots and patrol the perimeter fence with her. If Jamie doesn't like the boots they select, Jamie demands they choose another pair by shaking her head and pointing to the other boots. Over time, Jamie and Diana, and other staff and volunteers, have learned each other's cues, allowing them to communicate clearly with one another. Out of respect for Jamie's choices and the practical bounds to her freedom, Diana and others oblige Jamie's requests. They regularly patrol the perimeter with Jamie, sometimes up to fifteen times a day. While I was visiting, I grabbed some worn black boots Jamie selected and joined her and Diana. Jamie led the way as Diana wore a purple boot on her left foot and a tennis shoe on her right, while carrying a white boot in her free hand.

Diana's observations of Jamie remind me of a string of conversations I've had with my friend Scott Blais. When Scott was just fifteen years old, he landed a job as an elephant trainer at a safari park. Later, he worked in circuses and zoos. Like many other people, Scott was enamored with the idea of working with such beautiful, untamed creatures. But when he accepted his first job with elephants, he had no idea of what they had lost or how they were treated—or who they really were.

Kitty was one of the first elephants Scott worked with as a trainer. She had been isolated in a zoo for many years, and people also used her for riding. Elephants don't naturally let humans ride them. They must be "broken." Babies are torn away from their mothers and confined in a small cage or hole in the ground. They're starved, sleep deprived, and beaten into submission.

When Scott met Kitty, she was emaciated, antisocial, and aggressive. She tried to crush him while giving rides. To punish her, to show her who was in charge, the lead trainer told Kitty to lie down. He told Scott to strike her with a bull hook. The bull hook is one of the most notorious weapons used against elephants. It's a long, thick pole with a sharp metal hook on one end, and it's used on elephants' sensitive skin. Sometimes elephants are brought to their knees because the pain inflicted on them is so severe.

As a teenager, Scott did as he was told, despite his reservations. Though outsiders told him he was working with the best elephant trainers, he thought Kitty had a right to her aggression—to rebel against those who dominated her. After she was beaten, her chains were removed and she was taken to the yard and used for rides. From that time on, Scott never had any trouble with her. Her aggression—and her personality—became blunted.

After his experiences with Kitty, Scott tried to temper his treatment of the elephants. He began to see the physical and mental burden exacted on them. They slouched, limped, and cowered. They avoided eye contact and withheld their emotions.

Today, thousands of elephants live in captivity. Throughout the world, they are used as performers in circuses, exhibits in zoos, and actors in television and film. They are forced into labor and used as tourist ploys and ceremonial props.

Even when they aren't being threatened with a bull hook, elephants are subject to the whims of humans. They're kept chained in lonely quarters and transported in cramped train cars or tractor-trailers, many miles from their normal lives. Elephants naturally live in sub-Saharan Africa and South and Southeast Asia, and they possess some of the most intelligent brains on the planet. They converse through touch, visual displays, and sounds, including rumbles and infrasonic calls for long-distance communication. By stomping the earth's surface and triggering vibrations, they transfer information seismically. Elephants have a fission-fusion society—like humans and chimpanzees—in which multiple family groups come together and socialize. As a matriarchal society, elephant groups rely on their elder matrons for guidance and protection. They can live for up to seventy years in the wild and traverse long distances, sometimes up to dozens of miles a day. In the wild, they make countless daily decisions—about their own lives and the lives of their kin. These choices are taken away from them in captivity.

Within a few years of meeting Kitty, Scott met an elephant named Rasha who changed his life, and eventually the lives of other elephants. One day, Rasha resisted lying down for a bath. Just as Kitty was, elephants are often made to lie down before they are beaten. Another trainer suggested Scott softly talk with Rasha. He gently asked Rasha to help him, and she did. To Scott's surprise, she went even further. She gently wrapped her trunk around him and began a dialogue of squeaks and rumbles, all signs of trust, respect, and affection. She showed Scott how powerful his kind words and touch could be. Rasha—like Kitty and Jamie—merely wanted her choices to matter.

Scott's encounter with Rasha opened his eyes and heart. He wanted to leave behind a dominance philosophy that robbed the elephants of their freedom and sovereignty. The suffering he had witnessed in elephants reminded him too much of historical and modern-day human suffering. Gradually, as Scott changed his approach, Kitty also opened up to him with trust and love.

MODERN SLAVERY, WITHOUT THE CHAINS

The late civil rights activist, comedian, and author Dick Gregory wrote that when he saw animals like elephants in circuses, he thought of slavery: "Animals in circuses represent the domination and oppression we have fought against for so long. They wear the same chains and shackles."[1]

Gregory joined civil rights demonstrations in the early 1960s, including the March on Washington and the Selma to Montgomery March. He is generally considered the first African American stand-up comedian to make white audiences laugh at the absurdity of their own bigotry.

Though it is as old as the first city-state, the history of human slavery is one of our rawest realities. Soon after the dawn of civilization, the Egyptians kidnapped and enslaved people through expeditions down the Nile River, while the Romans began robbing thousands of people of their sovereignty through their military operations. Within the first millennium, slavery had become a normal practice in Europe, first in England's rural, agricultural economy, and later throughout the continent. Portuguese traders kidnapped and trafficked the first large group of enslaved humans, whom they called cargo, from West Africa to Europe by sea—establishing the Atlantic slave trade in 1444. Soon after, Spanish explorers took the first enslaved people from the African continent to what would become the United States.

It took more than two centuries—and many generations of enslavement, torture, and murder—for the Emancipation Proclamation to be issued by President Abraham Lincoln and for the passage of the Thirteenth Amendment to the Constitution outlawing slavery. But human slavery has not yet ended. Today, despite important successful abolition efforts, many men, women, and children remain enslaved. Researchers estimate that at least thirty million people are trapped in various forms of slavery throughout the world. India, China, and Russia account for more than half of all slavery, though it is found in nearly every country on the globe. Victims can be born into servitude, exploited in their homes, or transported to an exploitative situation. Today, most enslaved humans are not in chains, but they are subjugated and controlled by other means. They are forced to perform sex acts, domestic servitude, or hard labor, or to enlist in war as child soldiers. Many of the products we buy today come to us through slave labor, from animal products to diamonds, sugar, and gold. In all of these situations, people lose their sovereignty—the right to govern their own lives.[2]

Almost ten years ago, I met a woman named Lucee, who was applying for asylum in the United States. Her story of slavery is one of the worst I've ever heard. When we met one frigid evening in my clinic, I quietly listened and carefully documented her story and her scars. A West African rebel group targeted her hometown, kidnapped Lucee, mutilated her, and forced her into sexual slavery. At the time of her kidnapping, she was a young woman, almost a girl. Years after she was captured, a group of soldiers liberated her. To escape the terror of being captured again, she fled to the United States. As I wrote to the lawyer representing the US government, I've examined many individuals who have been tortured. Lucee's case was particularly compelling because of the intensity of suffering she endured. Before I met Lucee, I couldn't have imagined her story was possible. All her choices were taken away from her. Her perpetrators tried to reduce her to an object that didn't matter—an exhibit, a performer, a prop.

When I first met Lucee, I was concerned she would commit suicide. Pieces of her were already gone. For years, she lived with severe pain, depression, and PTSD. She cried every day. She couldn't sleep, and when she did, nightmares about her captors invaded her slumber. She felt ashamed and powerless. She struggled to hold on to hope. At the same time, she felt a strong need to control her environment.

After her release, she married, had children, and found work. When her husband became physically and emotionally abusive, she garnered the courage to leave him while she was pregnant with their child. She became independent and drew strength from her kids. As Lucee's sovereignty returned, so did she. She prevailed through a lengthy asylum application. Slowly she won back her life, her body, and her mind. When I testified in immigration court on her behalf, I was overwhelmed by her quiet fortitude and faith, even after years of migration and a long battle for refuge. Several years after I first met her, she was finally granted asylum.

Through Lucee, I discovered how living through the unimaginable is possible, and how each Phoenix rises on their own terms, some slowly, some more quickly. She was like a flame, almost extinguished but never fully dowsed. At times, when I looked at her, her face appeared blank, disguising a well of sadness underneath. But she was always in there—striving for another shot at unbound freedom and a time when her mind could escape its invisible cage.

Fortunately for many women, men, and children like Lucee, ending modern slavery has become a global priority. It falls within the United Nations Millennium Sustainable Development Goals, and groups like the Freedom Fund, a private philanthropic initiative, exist to end modern slavery. But even as we fight to end slavery, we must also consider what it takes for enslaved individuals to meet freedom and sovereignty at their door. Even if they escape slavery, cognitive, emotional, and social scars still hold them captive. They experience memory problems, intrusive flashbacks, anger, and guilt. How can individuals recover from such viciousness and pain? It's possible that at least some answers lie in our experiences with animals.

SOVEREIGN HEALING

Elephants were the first animals in whom mental disorders were widely publicized. In 2005, psychologist and ecologist Gay Bradshaw described PTSD in free-living elephants whose families had been hunted and killed in their native land. Soon after, she showed how elephants in close confinement also displayed Complex PTSD. The *New York Times Magazine* featured her findings in a 2006 article by Charles Siebert called "An Elephant Crackup?,"[3] and Bradshaw detailed them in her 2009 book *Elephants on the Edge.*[4]

Bradshaw and others have shown how elephants can recover from severe trauma, catalyzing our understanding of mental suffering and recovery in people and animals by spurring the field of trans-species psychology. As she and others have pointed out, our inferences about people and animals should be bidirectional—not unidirectional—if we really want to understand resilience.[5]

Even before Bradshaw published her findings, my friend Scott believed elephants could recover if given the chance. He quickly realized that elephants need enough space for sovereignty—where they can become independent again or for the first time in their lives.

In 1995, Scott cofounded the Tennessee Elephant Sanctuary. It is still the largest national habitat refuge created for African and Asian elephants in the United States, though Scott and his wife Kat, whom he met there, have since left the sanctuary.

Scott and Kat have shared many stories with me about how the elephants suffered before they arrived at the sanctuary. But it was their incredible resilience that captured my imagination.

The sanctuary's central philosophy revolves around restoration of the elephants' sense of agency.[6] There, they can exercise free will. The elephants wander where they want throughout verdant forest, in and out of the barn, to and from the pond. They choose their own food and decide whether to stay and visit with friends or seek solitude. They're empowered to make their own choices. They aren't punished for taking initiative—threatened or dominated—as they were in zoos and circuses.

For a few, recovery is almost immediate. Others take longer. Sissy—whom Scott and Kat describe as an old soul—was captured from the wilderness after she watched her family being shot. She was shipped to the United States in a box and kept in a zoo. While she was chained in a zoo barn, a flash flood swept her away. She was presumed dead until she was found with her trunk above water. Though she emerged with a paralyzed trunk, she was placed back in the barn, in chains, where she remained for years until she was sent to another Texas zoo. There, she sustained a gang beating by her trainers while chained. They wanted to make a demonstrational video for other zookeepers. One year later, the media covertly obtained a video of the beating. As a result, the city council voted unanimously to send her to the sanctuary. With time, she slowly found herself and developed loving relationships with Scott and her other caregivers. But they couldn't take the place of other elephants.

Over time, Sissy developed a deep friendship with another elephant named Winkie. They grew infinitely together. Though Winkie was traumatized when she arrived at the sanctuary, Scott stayed patient, relinquishing control to Winkie. Three days after her arrival, Scott was pressure-washing a stall when Winkie pressed her hind end close to the bars and began vocalizing, looking for reassurance. Scott reached over and touched her. She vocalized more. He touched her again, and she became even more vocal. He turned off the pressure washer and went over to rub her, and she rumbled. That sound—an elephant rumble—signals a huge emotional release.

In 2006, Winkie killed one of her caregivers, Joanna Burke (known as "Josie")—a thirty-six-year-old woman who had dedicated her life to people and animals.[7] Josie had walked around to Winkie's right side to look at a swollen eyelid thought to have been caused by an insect bite. During what has been described by some as a flashback, Winkie spun around, striking Josie across the chest and face. Josie fell backward and Winkie stepped on her, killing her instantly with thousands of pounds of pressure. Scott tried to save Josie, sustaining a broken ankle, as he kept saying Winkie's name over and over again until she snapped out of it.

Members of the public asked whether Winkie would be killed, but Josie's family supported the sanctuary's decision not to euthanize Winkie, saying Josie would not have wanted her to be harmed. Josie realized the risks of her work and had said as much in an earlier conversation with her parents. At her family's request, she was buried on the grounds of the elephant sanctuary. In lieu of flowers, her family asked that donations be made to the sanctuary. Their compassion seemed in keeping with Josie's life of kindness and service—from bringing cats home from the pound as a child, to tutoring children and adults in Appalachia. As she was laid to rest, Josie's brother asked if anyone had anything else to say. Her family and friends fell silent as a senior elephant trumpeted in the distance.[8]

After Josie's death, the Tennessee Wildlife Resources Agency and the Sheriff's Department investigated the incident, calling it an accident.[9] Though the sanctuary was in full compliance with all regulations, it subsequently changed its policies and required all caregivers to keep a barrier between them and the elephants.

After Josie died, Winkie became withdrawn. One day, Winkie was leaning against a bar and Scott went to comfort her. She vocalized. He believes she realized what she had done and deeply regretted it. At times, it felt like she was punishing herself by isolating herself. Moving forward, she grew. Though Scott still doesn't fully understand what happened, or if her reaction was related to a flashback of PTSD, he believes Winkie has always deserved continued love, respect, and sovereignty. With time, she has deepened her friendship with Sissy, often by venturing farther into the forest together.

A NEW ADVENTURE IN BRAZIL

In 2012, Scott and Kat launched Global Sanctuary for Elephants, which aims to create vast spaces around the world for captive elephants to recover from decades of physical and emotional trauma. After four years of gathering support and securing land appropriate for elephants, they started their first project in Brazil. Throughout South America, there are growing efforts to ban the use of elephants in entertainment, triggering their release from circuses and zoos. With almost three thousand acres, the sanctuary has plenty of space for all fifty captive elephants across the continent.

Maia and Guida were the first two elephants to join the Brazil sanctuary. Though little is known of their early lives, they were likely stolen as infants from a jungle in Asia. Shipped as cargo to South America, they were forced to participate in circus performances, very likely beaten and broken in the same way other captive elephants are. For decades, they were confined in small, bleak spaces.

Determined to understand their struggles and personalities in the months leading to their sanctuary arrival, Kat visited the farm in Paraguaçu, in eastern Brazil where they were kept. Beyond rows of coffee plantations, she came upon Maia and Guida in small pens separated by an electric fence. Kat and Scott learned about their lives through two men who had continued to care for them, even without pay. The girls had each been chained in their pens for five years. Kat sat for hours, gently communicating with Maia, gaining her trust, and documenting her observations as Maia's body gradually softened and relaxed, allowing her personality to peek through. She then moved on to Guida's pen, where she noted her small, underweight body and the way she compulsively waved her body and bobbed her head. Kat took note of her pain, as well as her hope. She noted that she was still there, holding on.

At the close of the dry season in 2016, my husband Nik and I visited the Brazil sanctuary. Though unplanned, we arrived the same day Maia and Guida did. After driving twenty miles on a red dirt road, past jungle forest stripped for corn and soy fields and two small wooden bridges reinforced with steel, we came upon the sanctuary. We arrived as night fell. It was lush with vegetation and devoid of artificial light. As we learned over the course of a week, the sanctuary occupied a diverse landscape of thick vegetation, palm trees, sandy soil, and rocky streams. The landscape was similar to the place of Maia and Guida's birth.

The small rural community in the western state of Mato Grosso rallied around Maia and Guida's arrival at the sanctuary. Along the main dirt road, children waited in the nearby village with bamboo bouquets as the two passed by in crates carrying them to their new home. After a three-day journey, they arrived late on an October day. The international media captured their first steps toward freedom. Cassia, a local volunteer, taped live video, which was broadcast around the world. In the background, you can hear her quietly crying, recording the first time Maia touched the ground, along with her whispers to Maia: "Careful . . ." and later, "She's so beautiful." As Maia stepped out of her crate into an indoor-outdoor care center, the video captured Scott saying, "Excellent, my dear . . . welcome home." With her long trunk, Maia picked up the fresh red dirt and poured it over her body, seeming to realize her new life had just begun. Soon after, Guida joined her. That night, Scott and Kat slept near them on hay bales, listening as the girls rumbled together.

By the second day, Maia and Guida had their first opportunity to venture into the outdoor enclosure together. Kat and Scott initially had concerns about how they might treat each other. They were told Maia was aggressive. In the past, she had knocked Guida to the ground.

With freedom and the beginning of a more sovereign life, Maia and Guida quickly revealed a different side of their relationship. Their gates opened to rumbles and trumpets. They grazed and foraged, free to seek companionship or privacy. On their first night together, they opted to sleep outside the care center together, with Kat and Scott nearby. Guida was the first to leave the barn and sleep under the stars, but Maia soon followed. They communicated throughout the night, and their delightful sounds continued throughout the week we were there. While Nik and I helped clean their care center area, I heard one of the sweetest sounds. I looked up and saw the girls in the field, trumpeting, rumbling, and ear flapping—all signs of joy. Though Maia was labeled as aggressive at the farm in Paraguaçu, she behaved very differently in her first week at the sanctuary. She wasn't aggressive toward Guida. They shared food, and

Maia even let Guida stand over her and rub against her. While Maia softened, Guida displayed an internal toughness. Even within the first week, her compulsive waving and bobbing diminished considerably. After we left, Kat shared a video of the girls in a torrential rainstorm. They trumpeted, danced, lay down, and wiggled around in the mud—forever free of the chains that had bound them for so long. Though they have a long journey ahead, they are poised to rise.

Scott and Kat continue to prepare for the arrival of other elephants. Step by step, they are expanding the area the elephants have to wander through and quell their beautiful minds. They try to reserve their expectations, giving the elephants what they need and letting them surprise them with who they are. As they've learned, through sanctuary they can offer captive elephants a degree of sovereignty. The elephants discover who they are, express themselves, and learn from their mistakes. They learn to control their own destinies.

It's difficult to know how much space elephants or other animals need to feel a sense of sovereignty. Scott has observed that each time the elephants moved, from a few acres to hundreds of acres to thousands of acres, they transformed even more than expected. He hopes to look up one day and see that they have disappeared into the forest, hidden to the human eye and influence. As he's learned over time, we really can't know them until they've had the opportunity to get to know themselves. And they can begin to know themselves only when they are given the sovereignty they need and deserve.

THE FOUNDATION FOR RECOVERY

When Scott and Kat first told me how elephants behave in circuses and zoos, much of what they described sounded like learned helplessness, a condition in which a person suffers from a sense of powerlessness. After persistent abuse, victims believe nothing they do matters, often leading to severe depression, as it did with Lucee. The brain learns that success is beyond its control. It can't affect the outcome. Once "conditioned" to this belief, hope is lost. In effect, victims learn to become helpless.

Martin Seligman and Steven Maier—experimental psychologists who drew upon Ivan Pavlov's classical conditioning experiments—first studied learned helplessness in dogs.[10] They tricked trapped dogs into thinking they could control electric shocks. But when the dogs realized they had no power to control the shocks, they stopped trying. They cowered and began to behave helplessly in other situations too. Soon, other scientists replicated their experiments using cats, fish, and rats. Some of the animals became physically ill and died. Later, they became interested in reversing the learned helplessness they had inflicted on dogs. So they started a new set of experiments, this time by giving dogs more control over their situations. The dogs didn't cower, and they became more resilient. Seligman went on to show how people could similarly overcome learned helplessness.

In 2002, findings from Seligman and Maier's experiments with dogs were used to craft enhanced interrogation techniques like waterboarding.[11] Two psychologists contracted by the CIA drew upon the foundational experiments to inform the architecture of its torture program and induce learned helplessness in human prisoners.

Just as Abraham Maslow suggested, self-determination is critical to well-being and resilience. In places of sanctuary, respect for personal sovereignty fosters the Phoenix Effect, perhaps because of changes in subcortical areas of the brain. Sadness, blunted personalities, maladaptive behaviors, and passivity fall away. Neurotransmitters in the brain involved in learned helplessness and related mental disorders aren't static. A sense of control—through empowerment—could actually change the amount and strength of nerve connections in the brain, including in the emotional centers like the amygdala or the hippocampus.[12] Neurogenesis, the birth of new brain cells, is also more likely to occur when inescapable sources of learned helplessness are removed. Though brain plasticity is suppressed during stress, a sense of control returns the brain to a healthier, more flexible state. In reality, learned helplessness isn't about helplessness: it's about freedom and independence. Autonomy is the enemy of helplessness, and respect for sovereignty fades the self-protective restraint tattooed on the mind by domination.

Today, we know that people and animals display varying degrees of independent decision-making skills. In recent decades, scientists and philosophers have shown how animals can act autonomously, like many humans.[13] Some have suggested that the capability for rational decision-making and making moral judgments should be used as strict criteria for whether an individual or species qualifies for legal rights[14]—even though these terms would exclude some humans. Though it's important to acknowledge capacities like autonomy and their implications, we already know through studies like Seligman and Maier's that freedom of choice is critically important to mental and physical health—whether in the case of a dog, a cat, a rat, a child, or a human adult with a normal or abnormal IQ. Given all we already know, why not respect the need for sovereignty now rather than waiting for even more evidence?

A UNIVERSAL COMMITMENT TO SOVEREIGNTY

Truly respecting the sovereignty of vulnerable people and animals is a significant challenge in our increasingly global society. It has become too cheap and easy to rely on the coerced sweat and tears of women, children, and men involuntarily co-opted into slavery and servitude. And—as in other areas of society where there are links between violence against people and animals—there are links here. Associations between the international wildlife trade, sexual slavery, and slave labor are but one example. The poaching, sale, and trade of animals used for food, medicine, commercial products, and entertainment, in developing and industrialized countries, fund and fuel the purchase of arms and ammunition, aiding sexual and other forms of human slavery.

How can we move forward in a world where there are so many connections between how we treat the sovereignty of people and animals?

Sue Donaldson, a Canadian author, and Will Kymlicka, her husband and a political philosopher, have suggested what some may see as a radical vision.[15] They oppose the idea that international justice is an exclusively human matter, or that animals' needs should be ignored in a human-animal society. They argue that domesticated animals should be considered citizens, wild animals should be granted something like national sovereignty, and liminal animals (e.g., pigeons, squirrels, and others who typically live among humans but aren't under our control) should be treated as denizens. Historically, denizens were foreigners granted certain rights in countries like Britain. In the case of animals living in the wild, some land would be generally off limits to humans, but even in the case of citizens and denizens, personal sovereignty would be respected and could be overruled only if it served animals' best interests.

Elsewhere, there have been calls for international animal ambassadors, including by Muhamed Sacirbey, former Bosnian foreign minister and ambassador to the United Nations. During and after the Bosnian War, Sacirbey advocated for victims of the Bosnian genocide. Since then, he has continued to work on human rights issues. On July 4, 2014 (US Independence Day), he penned an article for the *Huffington Post* entitled "Do Animals Need a UN Ambassador?"[16] Responding to his own question, he suggested that such an ambassador could speak on behalf of animals on poaching, inhumane treatment, and related issues. He proposed that we might even start with Special Representatives, Rapporteurs, and Experts, who are regularly appointed by the United Nations Secretary General to review and advise on global issues. Since crimes like the illicit wildlife trade often accompany severe violations of international law, like human slavery, Sacirbey argued that appointment of animal representation is already within the scope of the United Nations mission. He wrote that we should "undertake [an] effort to imprint a culture that respects animal life and [the] environment beyond its possible human exploitation," and that "such promotion of empathy would also enhance respect for human rights."[17]

International law is already beginning to reflect the reality of these suggestions. For example, in Zurich, Switzerland, a government-appointed advocate already represents the interests of animals in court. In 2014, the International Court of Justice at The Hague, the principal judicial organ of the United Nations, ruled against Japan's whaling program in the Antarctic.[18] In July 2016, the Ninth US Circuit Court of Appeals in San Francisco ruled that the navy could no longer use an ultra-loud sonar array in most of the world's oceans, in order to protect whales, dolphins, seals, and other animals at home in the sea.[19] Though these rulings don't exactly reflect what Donaldson, Kymlicka, or Sacirbey have suggested, they aren't too far off.

Imagine what Donaldson and Kymlicka's ideas or Sacirbey's suggestions would mean for chimpanzees like Jamie or elephants like Maia and Guida. How would their lives have been different? And what would such a deep respect for sovereignty mean to people like Lucee? What type of empathic culture would be imprinted upon us?

5 : LOVE AND TOLERANCE

COMBAT VETERANS AND WOLVES IN A DESERT FOREST

Love means more, since it includes fellowship in suffering, in joy, and in effort, but it shows the ethical only in a simile, although in a simile that is natural and profound.

— ALBERT SCHWEITZER

On a cool Southern California morning, my sister Carin and I drove ninety miles north of Los Angeles to the Lockwood Animal Rescue Center, home to the Warriors and Wolves project in the Los Padres National Forest. As we made our way north to the sanctuary, our ears gradually adjusted to a change in elevation while her car rushed through open mountain roads. The country music filtering through the radio slowly turned to static, and the changing terrain contrasted starkly with the bustling city we had just left. Concrete faded from our view, gradually replaced by fragile desert dotted with chaparral and pine trees.

We were en route to meet clinical psychologist Lorin Lindner and Desert Storm and Desert Shield war veteran Matt Simmons. Matt and Lorin cofounded the Warriors and Wolves project, which is available to combat veterans treated at the Los Angeles Veterans Administration Medical Center. The refuge is also home to wolves and wolfdogs they have rescued from injury or death. Like the combat veterans enrolled in the project, many of the animals come to the sanctuary traumatized and forgotten.

After about an hour of driving through sun-swept mountains, Carin and I turned off into the gravel parking lot of a local bar with natural brown wooden trim, and I dialed Lorin. She reiterated the directions to the refuge, and we turned out of the parking lot onto a main road, eventually turning onto a narrow, pale dirt road. When we came to a gate at a dead end, we called Lorin through the intercom system and she quickly greeted us in a mud-covered all-terrain vehicle filled with her dogs, including one pup called Cassidy with a missing leg.

We followed Lorin through a metal gate and headed toward a large shadowed barn where we met Matt and a few other dogs. I soon saw that the sanctuary serves many species—from humans and wolves to coyotes and horses. Before meeting the veterans and wolves, we spent time with a beautiful brown mare saved from the drug industry. Used for manufacturing drugs to treat symptoms of menopause, she had been repeatedly impregnated, hooked up to urine collection bags, and confined in a small stall where she was unable to turn around. Her prior life seemed a world away from the sanctuary, where she was able to nuzzle her foal, roam with other horses, and eat all the carrots she desired.

When we met the wolves, I was tickled by how much they looked and behaved like many dogs I know. Until recently, wolfdogs were called wolf hybrids. Wolves and dogs are close enough relatives that they can interbreed—begging the question of why we view and treat wolves so differently than we do dogs. Whereas dogs are known as "man's best friend," over hundreds of years wolves have become feared and demonized by prejudice and deceptive stereotypes.

Unlike the characteristic typecast, alpha male wolves aren't vicious. Carl Safina, a scientist who has observed wolves in their somewhat sovereign territory in Yellowstone National Park, points out that the leadership style of wolves doesn't rely on force and is not domineering or aggressive.[1] Instead, ranking male wolves are quietly confident, self-assured, and centered on calming others in their pack. They lead by example. They help raise, care for, and play with their young ones. As wolf researcher Rick McIntyre pointed out to Safina during his visit to Yellowstone, "Imagine two groups of the same kind—two wolf packs, two human tribes, whatever. Which group is more likely to better survive and reproduce: one whose members are more cooperative, more sharing, less violent with one another; or a group in which members are beating each other up and competing with one another?"[2]

Endangered by aerial hunts, human encroachment, and hunters' steel traps, wolves are increasingly under attack in their own homes. At the same time, more and more people are breeding wolves to dogs in captivity, creating untenable situations for these animals. Many illegally bred wolves are surrendered to shelters, where they are killed if they aren't rescued by organizations like the Warriors and Wolves project. Over the past several years, in addition to rescuing wolfdogs from shelters, Lorin and Matt have also spared wolves from backyard breeders and tourist parks. During one rescue operation at a roadside attraction in Alaska, Matt hiked through several feet of snow to pull a wolf out of a hole in the ground. When Matt found her, she was anchored to the icy earth with a heavy chain around her neck and buried under heavy snowfall.

As we walked around the sanctuary, our ears were filled with a melody of howls that sounded uncannily similar to my dogs' baying exchanges with coyotes at my home in New Mexico. The wolves and wolfdogs were different colors and sizes, with diverse personalities. Some had coats of white or medium to light gray fur. Others were a midnight black. Some were shy, others gregarious. We weren't cavalier about approaching them, but I was awed by how tolerant and gentle they were. They initially looked up at us with a rich curiosity but quickly lost interest in us in favor of playing with their wolf friends, Matt, and the other veterans. A few of the wolf ambassadors for the sanctuary joined us on our tour—including Huey, with his thin build, white and gray coat, and almost translucent golden eyes, and Wiley, with a darker shade of eyes and thicker coat.

As Lorin, Huey, and Wiley guided our tour of the sanctuary, we met veterans caring for the animals. The veterans in the program are men and women who have served in the air force, army, coast guard, marines, and navy. Some veterans come to volunteer, others are paid, and some come for the support. Many are still recovering from PTSD, traumatic brain injuries, and depression after returning from war, and conventional medical and psychiatric treatments have largely failed them. Like other veterans, they struggle to fit into society. They suffer from what was once known as shell shock or battle fatigue and the high emotional toll exacted by widespread death and destruction.

Combat, and the killing that lies at the heart of it, is almost indescribably traumatic—altering brain chemistry, marring the body, and twisting the soul. Warriors are witness to some of the most devastating losses, and often subject to complete or partial physical and emotional demise. In addition to risking their own lives, and witnessing the maiming and killing of their brethren, many soldiers must also overcome a natural resistance to hurt or kill others. Though it's too seldom discussed, many veterans wrestle with the psychological ramifications of killing others, even decades after returning from combat. They endure the wounds of war many of us willfully ignore. Despite being celebrated for their military service, their subsequent vulnerability to physical and mental pain is frequently viewed as an unwelcome consequence of war that few of us are willing to confront. As with other forms of vulnerability, it brings our shared culpability to the surface—causing many of us to turn our heads away in obstinate ignorance.

Constant fear, physiological arousal, and the terror of war regularly leave veterans with a disordered fight-or-flight response, resulting in psychiatric disorders like posttraumatic stress and depression. Soldiers are far more likely to suffer psychiatric consequences than physical disability or death. Between October 2001 and June 2012, nearly one-third of Afghanistan and Iraq War veterans treated at Veterans Administration hospitals and clinics were diagnosed with PTSD.[3]

Lorin, who still holds a full-time job caring for children and adults with mental health issues, speaks candidly about the challenges she has faced in her almost-thirty-year-long career of working with combat veterans. When Lorin told me about her difficulties treating veterans, I understood. As I've learned through my own patients, veterans with posttraumatic symptoms often struggle with an inability to trust, build relationships, or feel safe. Veterans are frequently plagued by guilt over what they have seen, done, or failed to do—even when their guilt is irrational. Many struggle with flashbacks, nightmares, and panic attacks, despite trying to avoid reminders of their trauma. One of my patients, a Vietnam War veteran who grappled with homelessness and addiction even decades after returning to the United States, would frequently awaken in the middle of the night crouched in a corner much like he did in the Vietnam jungle. He could never figure out how he ended up in that position, and it often took him hours to return to reality. Another veteran once told me that his triggers could take any form—an odor, a laugh, a voice, an accent, a word, a color, or certain tools or textures. It terrified him that he couldn't predict when, where, or how he would react. Even seeing his own expression in the mirror sometimes threw him back into a trap fueled by fear and anger, causing him to retreat into a silent hole. He was crippled by negative thoughts about himself and social detachment from his family and friends. Both men struggled with drugs and alcohol. Difficulty controlling their emotions and angry outbursts led them to numb their feelings with even more drugs, alcohol, and isolation.

At the Warriors and Wolves project, the veterans find something very different—a chance to rebuild trust, form important bonds with the wolves and other veterans, and become part of something greater than themselves. Though the sanctuary clearly acknowledges the importance of freedom and sovereignty to the veterans and wolves, the love they're shown is equally important. The veterans have an opportunity to rise from the ashes of war. Some of the most vocal supporters of the sanctuary are those who have overcome seclusion and homelessness. They have transformed physically and mentally, in part through commiserating with wolves who have also suffered bodily and emotional injuries. For some veterans, the refuge has offered a reprieve from spiraling into alcohol and drug dependence. They find deep meaning in the unconditional love and acceptance offered by the wolves, and what they can offer the wolves in turn. One of the war veterans, Jim, found that no orthodox therapies helped him when he returned home after ten years of military life. He was in the midst of drinking himself to death—what he calls the last chapter of his life—when he found the Warriors and Wolves program. He found solidarity and peace with the wolves: acceptance as part of their pack.[4] Fortunately, like the veterans, the wolves at the sanctuary are resilient Phoenixes. They slowly recover from surgeries to remove chains embedded in their necks, limps associated with being chained and confined, and their psychological wounds.

Some of the veterans have even made efforts to repay the wolves by protecting their kin. After learning about planned hunts in the northern part of the country, the veterans camp out in the wilderness, wait for hunters, and try to stop them from killing wolves. They share how the wolves have helped them through their struggles. In spite of all the dissociation, detachment, and misery many of them have suffered through, the warriors and wolves eventually find solace and healing with one another. Their ability to overcome severe trauma together is likely fostered by empathy, an evolutionarily ancient social quality that also places us at risk for emotional trauma.

The success of the Warriors and Wolves program parallels the growing number of similar programs that pair homeless dogs with veterans. Many of the foster dogs in these programs have been severely neglected and abused. Nonetheless, they serve as catalysts for love for veterans returning home from war. They offer protection from apathy, addiction, violence against others, and suicide.

RISING TOGETHER WITH UNCONDITIONAL "WUV"

Matt and Lorin first met after she launched an innovative program working with parrots and veterans. Lorin developed an affinity for animals in her childhood. One night before Thanksgiving, her mother walked into their kitchen to discover a seven-year-old Lorin cradling the defrosting body of a turkey. Later in life, in pursuit of a career in psychology, Lorin got to know animals living in laboratories. She would stay late after researchers left the laboratory to spend extra time with the animals. But she could never come to terms with what she calls the "collective trauma of all those lives."[5] She describes being baffled by a paradox in which animals were used in experiments, since they share our need for love and affection, with little to no ethical attention to the ways in which they suffer. She couldn't draw a clear line between the empathy she had for people and the empathy she had for animals.

While living in Los Angeles, Lorin became friends with two salmon-crested cockatoos—one called Salmon, or "Sam" for short, and another named Mango, or "Manny."[6] Both were taken from their homes in the Moluccan Islands of Indonesia and lived in at least ten homes before Lorin rescued them. She and the birds developed a fondness and love for each other that slowly influenced the untraditional path Lorin would eventually take as a clinical psychologist.

Through Manny and Sam, Lorin learned about birds' capacity for love. When Manny groomed Sam, Sam would often look up and say, "I love you, Manny." Lorin interpreted these vocal and other behavioral interactions as affection—just like the love she felt for them. As a psychologist, Lorin also understood the science behind their mutual love and affection. People and animals share brain structures and physiology that influence our aptitude for love. Connections within the brain make it possible for us to experience fear, rage, and aggression, as well as friendship and love. Though no single structure is responsible for the complex emotions associated with love, it is regulated through small structures deep within the brain called the limbic system, made in part by that little almond-shaped structure considered the center of emotion, the amygdala. It scans each experience for safety or danger and pleasure or pain, and then links each experience with an emotional flag that can help us make decisions in the future. By connecting with the hippocampus, it helps store, retrieve, and even reinterpret memories. Intense, loving experiences increase the likelihood that we will form strong emotional bonds and memories. These experiences also help our malleable brains recover from severe trauma.

The same year Lorin became the clinical director of a large drug and alcohol rehabilitation program at the West Los Angeles Veterans Administration Medical Center, she also started a parrot sanctuary, where she brought Sam and Manny to live with other birds. Parrots are often dismissed because of their small "bird brains." However, studies have shown that parrots have complex problem-solving skills and a keen sense of self-awareness, similar to great apes like humans and chimpanzees. These similarities are likely due to a phenomenon called convergent evolution, when analogous features, like the wings of birds and bats, emerge independently in species with different lineages. Though birds' brains are small, they are powerful, much like the seemingly unlikely power of a small, handheld smartphone.

Like humans and many other animals, parrots need to belong to a social circle. They are hardwired to be in contact with one another, in some way, at all times. In the wild, they live together in flocks. They call each other by name, rely on their parents and other flock members for an extended developmental period, and learn within a social context. Parrots have a large collection of social neurons that allow them to form close bonds that seemingly mirror the emotions and thoughts of others—a level of empathy so deep it's difficult to characterize. Unfortunately, most captive companion parrots live without other parrots, and even after they bond with humans they are often robbed of these connections, too.

Birds have become some of the most popular companion animals in the country, and they are typically either captured in the wild or bred, sometimes in mass-breeding facilities resembling puppy mills.

Captive parrots often experience unstable lives of constant rehoming, repeatedly forced to break and re-form bonds with people. In many cases, parrots are kept alone, resulting in a long solitary confinement sentence for birds who can live for up to a century. Like other social animals, parrots can develop psychiatric disorders like PTSD, depression, and compulsive disorders. Parrots affected by mental disorders routinely self-mutilate, pace, and scream. They can also develop sleeping and eating disorders.[7]

Soon after she established the parrot sanctuary, Lorin started bringing homeless veterans from her drug and alcohol recovery program to the refuge. There, she discovered something profound. Some of the seemingly unreachable, stoic veterans wracked with PTSD and addiction began to open up to the birds. Many were taken in by Manny's calls of "Hewwo, I wuv you."[8] Ultimately, Lorin's unorthodox approach led to the construction of a sanctuary filled with trees, fountains, and flowers—right on the Veterans Administration Medical Center grounds. The sanctuary is aptly named Serenity Park. There, veterans receive mental health care, overcome addiction, and gather resilience, while also helping parrots rescued from abuse and isolation. With their intense sensitivity, the parrots are able to connect with the veterans, without judgment, drawing them out of their deep-seated pain. Some researchers believe the veterans are able to rewire their brains—creating new cognitive and emotional pathways around damaged brain tissue, much like a heart can develop new blood vessel circuits around scarred cardiac tissue. And as the veterans heal, the parrots literally mirror their recovery, perhaps because of the parrots' large number of social neurons and deep empathic responses. As a result, the parrots seldom rock back and forth or pace with frenzy. They become less aggressive and stop mutilating and starving themselves. And like the veterans, they are better able to sleep.[9]

The results seen at Serenity Park and the Warriors and Wolves program should not be surprising. Survivors who have lived through war, torture, and genocide are more likely to recover by bonding with others.[10] Even fleeting love and the love of strangers can help us thrive. To prove this point, researchers tested college students early in 2001 and again in the weeks following the terrorist attack on the World Trade Center on September 11, 2001.[11] Love—even in fleeting doses—fueled resilience and protected students from sinking into depression. This pattern is seen throughout human society, and among animals.

Sharing love with animals is similarly beneficial. People living with companion animals are generally healthier and happier, with lower blood pressure, lower cholesterol, and a reduced risk for dying from a heart attack.[12] Being involved in caring relationships with animals raises levels of the brain chemicals serotonin and dopamine, lowering the risk for severe depression and other mental illnesses.

In a touch of irony, Serenity Park is also where Lorin and Matt found love. Before creating the Warriors and Wolves project with Lorin, Matt struggled with his own pain. He has spoken publicly about his battle. He returned from the war in Iraq to his hometown in Ohio, where he had been a high school and college All-American football star. Matt suffered from PTSD, and he self-medicated with drugs and alcohol. After traditional therapies didn't work, he was referred to Serenity Park. Slowly, Matt healed and ended his dependence on drug use. Through their work together, Lorin and Matt fell in love, and they were married in 2009 by actor James Cromwell.

WAITING FOR LOVE

Before I first visited Lorin and Matt, I learned about the plight of parrots through my friend Wes, who has a parrot sanctuary in Canada. Many of the birds who are now in his care were originally purchased by people who later discovered that they were incapable of dealing with the parrots' complexities. Before joining Wes's sanctuary, some were even shoved into drawers and darkness, much like human prisoners. Many of the birds have come to him with blatant physical and behavioral signs of suffering—bodies plucked free of feathers, screaming fits, attacks on other birds and people, and complete social dysfunction. Over time, Wes and his wife have rescued dozens of parrots, all while maintaining their full-time jobs.

One day, Wes and I were talking about his childhood. He paused, looked down, took a breath, then looked me in the eye and said, "I was alone and hungry most of the time growing up. I was rejected twice by my mother." When Wes was between the ages of seven and nine years old, his mother sent him to live in a foster home on two different occasions. He was also sick a lot as a child. He can still vividly recall being in the hospital for a couple of months, looking out the window, and waiting for his mom to visit when she said she would. But she never showed up. After he told me this story, he quickly followed with "This might be why I understand what it means for a parrot to wait. To wait to be heard, acknowledged, to feel like you matter." It was difficult to picture Wes, who is now a tall, solid man with a bald head and clear blue eyes, as the child he described. But it helped me understand his love and devotion to the birds, as well as how his empathic relationship with the parrots has helped fuel his and their recoveries.

The science of love has a long, twisted and, at certain times, demented history. During the first half of the twentieth century, many psychologists and psychiatrists suggested that affection should be withheld from children. Fortunately, by the latter half of the twentieth century, pediatrician Benjamin Spock, along with British psychiatrist John Bowlby and developmental psychologist Mary Ainsworth, who had both yearned for the affection of their own mothers, had reversed this notion. They pointed out that, without love and affection, children have more difficulties coping with stress and adversity throughout childhood, adolescence, and even adulthood. Around the same time, scientist Harry Harlow began his series of controversial maternal deprivation experiments—those that showed that, like human children, baby monkeys desperately need the love of their mothers.

Scientists have since learned that children respond to the presence or absence of love through what is called the hypothalamic-pituitary-adrenal axis, which helps regulate the body's response to stress through the cortisol hormone. This phenomenon is seen in people and animals.[13] Caregivers are central to whether a child can rise from adversity as an adult. Animal parents like mice and rats help their pups in similar ways, and the same is true for monkeys, birds, wolves, and other animals. Young ones deprived of affectionate care are at increased risk for psychiatric disorders like depression and PTSD, since their brains cannot adequately respond to traumatic situations. In contrast, children raised in supportive, nurturing environments are more likely to be resilient in the face of trauma. This tenet holds true even during war.

Parental love can actually increase the level of resilience of children caught in conflict zones.[14] Love and empathy are at the heart of development, growth, and the ability to recover from trauma.

CONTAGIOUS COMPASSION

Loving others, even those we don't know personally, is good for us. Empathy—the ability to experience or understand the emotional feelings of another—and altruistic love can add years to our lives and increase resilience.[15] Helping others actually activates the same parts of the brain lit up by rewards and pleasure.

Our need to love others extends well beyond our close circle of friends and family. Love is not merely the opposite of hate, but also a refusal to remain indifferent. Many of the torture survivors I have met embody this idea. Doc refused to remain indifferent to injustices around him. Though he was tortured as a result of his efforts to end injustices committed against other people, his empathy for others helped fuel his recovery—much like Matt's empathy for other veterans and wolves helped drive his ability to recover from the trauma of war.

Many animals display empathy and altruism.[16] Chimpanzees console each other when they are distressed. Monkeys take extra care with physically challenged monkeys in their group—they will even starve themselves to prevent another monkey from being shocked with electricity. Similarly, rats risk their own lives to free others by breaking into their cages. Mice also recognize when their fellow mice are suffering and sacrifice their own needs to reduce their friends' suffering. And the list goes on.

It makes sense that the capacity for empathy is spread widely across the animal kingdom. Empathy promotes peace, prevents violent escalations, and preserves life. Even science supports the idea that empathy, compassion, and tolerance—not aggression—are the most effective antidotes to violence. Known for coining the term "moral molecule" to describe the hormone oxytocin, scientist Paul J. Zak has studied the chemical basis for kindness. He and others have shown how a large number of caring, compassionate actions among people and animals cause oxytocin release,[17] which calms fear and aggression and often leads to more caring interactions. These studies show how kindness, empathy, and compassion are contagious. Perhaps these virtues could even help stop the contagion of violence.

LET LOVE LEAD THE WAY

The ubiquity of violence in our society is, at times, paralyzing. Violence creeps into our homes, our workplaces, and even our schools. But even amid extreme violence and despair, some people remind us that it is possible to find the strength and courage to respond with forgiveness and hope. The veterans at the Warriors and Wolves project show how it is possible to turn from a violent environment to love, compassion, and nonviolence. And a mother, who lost her son to a mass shooter at Sandy Hook Elementary School in Connecticut, is another one of those people.

On a Friday morning in December 2012, an armed twenty-year-old named Adam Lanza entered an elementary school in the quiet town of Newtown, Connecticut. After killing his mother, he took three guns from his house and went to the school wearing black fatigues and a military vest. Even though the principal of the small school had installed a new security system, he was able to force his way into the school with a semiautomatic rifle and two pistols. About seven hundred children were in their classrooms, and by the time he committed suicide, he had killed twenty first graders and six adults trying to protect their students.

Many people have tried to understand what happened. In 2014, Connecticut's Office of the Child Advocate released a one-hundred-fourteen-page report in search of an answer.[18] As the report notes, earlier in Lanza's life he was diagnosed with autism spectrum disorder, a sensory integration disorder, anxiety, and obsessive-compulsive disorder. But most people living with these disorders do not commit violent crimes. At the time of his death, according to a report issued by the Office of the Child Advocate, the Office of the Chief Medical Examiner deemed he was anorexic, with evidence of malnutrition and brain damage. He had also become severely socially isolated and preoccupied with guns, violence, and mass murder—with many missed opportunities for intervention.

Just twenty-one months after her own autistic six-year-old son was killed at Sandy Hook Elementary School, Nicole Hockley spoke about her loss at a conference on violence prevention, where I also talked about my human rights work.[19] With tremendous courage, she described learning that her son Dylan had been killed in the mass shooting. Hours earlier, she learned that her other son had survived. But rather than responding with malice, she asked us to "let love lead the way." She stressed the need for empathy, compassion, and connectivity.

In closing her talk, she told a story about Dylan. She described how he would flap his arms like a butterfly, and she alluded to the butterfly effect, the idea that a single occurrence, no matter how small, can alter the course of the universe forever. She voiced her hope that his memory would lead to significant change, much like one small butterfly can influence a hurricane on the other side of the world. In honor of Dylan's memory, she and her husband have since created a foundation to help other children with special needs achieve their full potential. And, with other parents, she helped create Sandy Hook Promise, which aims to prevent gun-related deaths.

About a year after I attended the conference where Hockley spoke, I learned about a certified therapy dog named Juno, after the Roman goddess best known for her protective proclivities. Juno's life started in a laboratory research facility, where she was used in experiments. In April 2010 the laboratory went bankrupt and closed their doors, and Juno and more than a hundred other beagles were left to die. The dogs were rescued on Independence Day in 2010, and a woman named Kate Aubry later adopted Juno.

On New Year's Eve in 2012, Kate received a call requesting that she and Juno come to Newtown to meet with some of the children affected by the shooting at Sandy Hook Elementary School. Kate told me she immediately said yes. As soon as Kate and Juno arrived at the school, the school doors opened. Hundreds of kids poured out toward Juno. She swiftly acclimated to the kids, and a sea of children's hands quickly covered her belly. One girl asked Kate if Juno could come to her classroom. After receiving approval from the teacher, Kate and Juno headed to the girl's classroom. While there, Kate told me that another girl came over and hugged Juno like her life depended on it. Juno just let the girl hug her as long as she needed to. The girl then packed her bag and left. Afterward, the teacher told Kate it was the girl's first day back at school after losing her brother in the school shooting. Kate cried the whole way home and decided she had to do better by Juno and dogs like her. Juno became the inspiration for the BeFreegle Foundation, which now places in permanent homes dogs who might otherwise be killed at the end of laboratory experiments.

The love led by these remarkable people suggests something profound. Whereas an absence of love—in the form of indifference, neglect, abuse, or hate—can fuel rage and violence on an individual and global scale, leading with love can help us heal. Given its contagious properties, it could even protect against some of the most severe forms of violence—the torture Doc suffered, the war Matt fought in, the mass shootings too many families endure. And some people, like my patient Benjamin, have suggested that it might even protect us from mass atrocities like genocide.

Benjamin and I met in my Washington, DC, clinic for torture survivors. He was a short, serious man with thick bifocals. Deliberate with his speech, he chose his words carefully and refused to waste a single word. As I learned when I evaluated him for traces of torture, Benjamin lost almost all of his family during and after the Rwandan genocide. As often happens in warfare, propaganda was used to degrade victims of the genocide. In 1994, after a campaign of hateful rhetoric, Hutu extremists targeted and killed Tutsis and moderate Hutus—even as they fled to churches, where they were slaughtered with machetes and burned alive. Over approximately one hundred days, eight hundred thousand men, women, and children were killed. Identified as Tutsis, Benjamin's family was among those killed. But the violence against them did not stop there. Benjamin's only remaining parent, who was identified as Hutu, was killed in retaliation for the genocide. Afterward, Benjamin became politically active with groups demanding democracy and transparency in Rwanda. Because of his activism, he was targeted and tortured.

Despite all that happened to him, Benjamin still loves his country and his countrymen. He believes the only way forward is to live as a tolerant community of brothers and sisters. As he pointed out to me, one of his first lessons in life, as a child of mixed ancestry, was that love should know no boundaries.

After the Rwandan genocide, Nelson Mandela issued a plea to South Africans to take two roads named "Goodness and Forgiveness." As president, he established the Truth and Reconciliation Commission, which emphasized forgiveness and tolerance. His approach is believed to have helped South Africans avoid further mass atrocities like genocide even after years of oppression and division under apartheid.

The American poet Maya Angelou once wrote that "love costs us all we are . . . yet it is only love which sets us free."[20] If scientists are right, love pays off far more than it costs us, protecting us from disease and disorder. But it could also save us as a society by calming aggression, preventing destruction, and fostering hope. Whereas torture, war, and other forms of violence foster an illusion that the social order can be broken and remade at will—destabilizing society and becoming a cancer that metastasizes uncontrollably—love can halt that cancer at its door. It can help us create more sanctuary in a wounded world.

If we look and listen closely, we can learn a lot from combat veterans who have turned away from war, grieving mothers who have left behind malice, and torture survivors who have rejected hate and ignorance—instead turning toward love and tolerance. No matter how much sorrow there is, the light of love can peek through the darkest places. Just as Lorin and Matt have created space in a desert forest for the battered and beleaguered to rise from conflict and indifference, we could do the same in places currently plagued by antipathy. In doing so, we could help more individuals overcome hopelessness and despair.

As Nicole Hockley pointed out, we can each begin with one small, compassionate act that becomes many, perhaps altering the course of the universe forever.

6 : JUSTICE

SHELTER FOR HOMELESS CHILDREN AND THEIR COMPANION ANIMALS

Justice is the first virtue of social institutions, as truth is of systems of thought.
— JOHN RAWLS, *A Theory of Justice*

Love powers Matt and Lorin's efforts, but the pursuit of justice drives them. Like many other creators of sanctuary, they have a strong sense of social justice—a notion that philosophers and theologians have struggled for millennia to define. One of the most prolific writers on the subject of justice is philosopher John Rawls. In the latter part of the twentieth century, Rawls articulated justice as fairness, and he laid out two important principles of social justice.[1] The first guarantees liberty, whereas the second ensures fair equality of opportunities and consequences and that the greatest benefits and protections should be given to the most vulnerable members of society. Both principles are prefaced on the idea that each individual is impartially situated as an equal behind a "veil of ignorance," regardless of personal or social characteristics.[2] At the time, Rawls did not really address how animals fit into these ideas.

At its core, justice is a bridge between all other needs and rights in our relationships with others. For survivors, justice takes on a very specific meaning: that their lives will be treated as equally as the next. They know what justice is by virtue of what it is not. Without it, torture survivors Doc and Grace would still be deprived of freedom and the love they share with their families and friends. And though it's difficult to imagine the injustices they've overcome, it may be even harder to come to terms with the wrongs children endure.

In the middle of autumn, in the same Washington, DC, clinic room where I had examined Doc a year before, I met Sofia. She timidly walked toward me, accompanied by her attorney and guardian. She was on the cusp of adulthood, though her pain began as a young child in her own home. Her mother abandoned her, leaving her with relatives who repeatedly physically and sexually abused her. Throughout her young life she tried, often in vain, to shield from abuse her siblings and the animals she cherished. As a teenager she sought help, but the authorities didn't believe her. Fearing she had no other choice, she fled her home in Central America to seek asylum in the United States, where she hoped to also find a safe place for her siblings. In defiance of all she had lived through, Sofia's courage was striking.

When I met Sofia, she spoke in soft, brief sentences. She hesitated to describe the abuse she and her brother and sister had suffered. Her scars told many a story. Like other young people in her position, Sofia felt scared and alone. She was plagued with constant worry and frequent nightmares. While awake, she still heard her sister's cries. She perceived threats where there were none—for example, flinching when a classmate raised her hand, for fear of being hit. Nonetheless, she remained resilient. She thrived in school and dreamed of becoming an electrical engineer.

Children and adolescents experience trauma differently than adults do, compounded by their developing minds and the gravity of their vulnerability in society. Their wounds often extend into adulthood, carrying potential ramifications for their children and the world around them. Some children show few effects, while others suffer from long-term debilitating challenges. But like Sofia, they can be resilient—particularly when they are swathed in love and, through justice, can reclaim their sovereignty and dreams.

Sofia's story is far too common. As in other areas of the world, at least one-quarter of American children suffer from one or more forms of abuse. In the United States, about four children die each day as a result of neglect or abuse.[3] In these cases, a child's vulnerability is treated as a target, rather than the shield that justice demands. Their lives and deaths raise the question: How can we help child survivors rise from adversity and create a more just world for them and for future generations?

FROM OUTRAGE TO COURAGE

A few years after I met Sofia, I took a concrete highway in Oklahoma, where I had grown up, to Sisu Youth, a shelter for abused and homeless youth in the Oklahoma City metro area. Thick, muggy air filled the car as I rolled down the windows. Champ, my ten-year-old shepherd-husky dog who has since died, sat in the back and leaned his head out the cracked window to breathe in the scent of fresh alfalfa.

We found the shelter among some small storefronts near the Oklahoma City arts district and pulled into the parking lot of a red-brick church. As we got out of the car, Penny Reynolds, who founded Sisu Youth with her partner, Amber Richardson, moved purposefully across the hot pavement in her black combat boots. She bent down to greet Champ before engulfing me in a hug. We were quickly consumed by her magnetic presence. Large, dark-rimmed glasses framed her kind brown eyes and wrapped around her short, spiky brown hair. Following her into the church, we took an elevator to the lower level. Vibrant colors covered the walls, windows, and comforters, and warm, welcoming messages flanked the kids' lockers. Like many places I've worked in Africa, the facility's modesty hid its quiet power.

Penny first experienced homelessness at the age of nineteen. Like many other adolescents who become homeless, she struggled as a child. Early in life she took on adult responsibilities, caring for her mother through manic, depressive, and epileptic episodes. She moved from house to house and relied on friends at school for lunch money. She spent sixth grade hiding in the woods, seventh and eighth grade feigning illness and avoiding school, and her high school years skating by on D's and F's. By the time she graduated, she was picking up odd jobs and staying on her friends' couches. Soon she turned to living in her car. That time in her life left what she calls a permanent mark on her soul. After years of struggling with homelessness, Penny left Oklahoma to work on the railroad that runs through the southwestern states. A few years later, she returned to adopt two of her young cousins, now her sons, after they were removed from their biological parents. Penny was their only option outside foster care. She later met and fell in love with Amber, and the two of them began to blend a modern family.

Together, Penny and Amber—who measures about a foot shorter than Penny—have fostered and adopted five children. All of their kids had been removed from previous homes because of neglect or abuse. With more than three jobs between them, they've raised five kids on a limited income while Amber earned a master's degree in social work. Their lives are simultaneously ordinary and extraordinary—filled with homework, takeout menus, strict bedtimes, and minivans. Their sweet laughter defies a quiet fury sparked by the injustices their children and others face. Their outrage in an unjust world fuels their steady courage.

Even after their kids filled their house with giggles, tears, and tantrums, Penny and Amber couldn't forget how many other children remain in limbo. Reality defies Rawls's principles of justice—especially the notion that the greatest protections should be given to the most vulnerable members of society. Aiming to create a safe place where homeless young people like their own children and Sofia could thrive, they launched a movement called Sisu Youth. Loosely translated, "Sisu" is a Finnish word that means succeeding against all odds; overcoming extreme adversity with tenacity.

Every year, an estimated two and a half million children are without a permanent home, a number that is steadily rising.[4] Oklahoma ranks fifth worst among states in the percentage of children under eighteen who are homeless—more than forty thousand children. In Oklahoma, like elsewhere, pregnant, gay, lesbian, bisexual, and transgender youth and young people of color are disproportionately affected. Most homeless youth are between the ages of fifteen and seventeen.

Young people often become homeless while fleeing rejection, neglect, or the next traumatic violation of their bodies. In many cases, they have already endured years of abuse. In a study by the US Department of Health and Human Services, more than half of young people interviewed at shelters reported that their parents told them to leave or knew they were leaving and didn't care.[5] Another study by the agency revealed that about half had been physically or sexually abused.[6]

Adolescents who live alone on the streets are particularly susceptible to harm. Housing, employment, and other social and economic policies make it difficult for them to earn enough money for food, clothing, or shelter. Consequently, they are at greater risk for sexual exploitation, contracting diseases like HIV/AIDS, and being killed. Some stay awake all night in an attempt to stay safe while racked with crippling anxiety, depression, and PTSD.[7] Thousands of homeless youth die each year as a result of physical or sexual assault, acute or chronic illness, or suicide.[8] If they stay alive, they frequently remain on the fringes of society—unprotected and invisible.

As a doctor in New York and Washington, DC, I cared for many people who struggled with temporary and chronic homelessness. Though I did my best to treat their medical and mental illnesses, I realized my efforts would never be sufficient without addressing the root of their problems. Frederick, a young man in his late teens, was but one example. When I met him, he was sleeping on the steps of a church. He often came to see me for inhalers for his chronic lung disease, or for antibiotics for a lung infection. Because of paranoia related to violence he had suffered in the past, he refused to stay in a shelter. He required a home where he could feel safe, as well as steady support and unconditional love. He needed the structural violence perpetuated against him on a daily basis to stop. But it didn't, and he wasn't alone. Through my patients, I learned that anyone can become homeless, particularly individuals plagued by a history of violence. It's a form of vulnerability we often prefer to ignore when we pass people sleeping on sidewalks, panhandling on highway medians, or camping in the woods. But Penny and Amber opted for a different response.

Amid a background of desperation, and Penny and Amber's courage, Sisu Youth was born. In 2010, they started with a simple mission: to collect and share emergency resources for at-risk youth. Quickly they realized how little help was available for homeless and abused young people, so they launched a statewide campaign to establish a network of services for them.

After three years of research and advocacy, Penny realized there was a serious gap in assistance for struggling children and adolescents. In March 2014, she shared her concerns with an Oklahoma businessman who surprised her with a one-year/one-dollar lease for one of his vacant suites. Within two months, the organization was up and running with a day center—providing homeless young people access to a kitchen, library, clothing closet, and media room, as well as case managers, counselors, and tutors. Volunteers lined up to help thousands of teens.

As Penny and Amber soon realized, the resources the day center provided weren't enough. Many kids still needed safe shelter at night. At the time, Penny was trying to solve each problem that came to her attention with a series of phone calls, online pleas, and in-depth study of the law. One day, Penny worked feverishly to find a place for a sixteen-year-old pregnant girl to stay. She had been thrown out of her house and, because of her age, she couldn't rent an apartment or motel room. For Penny, it was a reminder of how many children are stuck in unsafe environments.

By the summer of 2015, Sisu Youth had opened a new facility in the lower level of the church we visited to offer temporary overnight emergency shelter and day center services for teenagers, many of whom don't qualify for child welfare services. Working with community partners, Penny helped establish Teen Safe OKC, a teen-focused community resource that connects young people experiencing hunger, homelessness, and crisis with safe spaces and resources in their local areas. The teens also began helping each other, serving as support systems for those who needed it most. Realizing the importance of policy change, Penny began meeting with members of the Oklahoma legislature. Initially intimidated by the prospect of congressional meetings, within the first few years she had already triggered legislative changes that could impact an untold number of kids.

THE NEXT CHALLENGE

When I visited Penny, she was trying to find a way to accommodate teens with companion animals in need of shelter. An estimated five to ten percent of people experiencing homelessness share their lives with companion animals, though this number is likely higher among youth.[9] Many homeless individuals refuse to enter shelters without their animals—perhaps driven by their own sense of justice. Homeless teens with animals also frequently struggle to find food and health facilities willing to accommodate them, deepening their marginalization. Although animals often offer protection for young people, they must also protect their animals.[10]

Studies show that homeless teens living with animals have lower rates of depression than their counterparts without animals. Just like Kate Aubry and her rescued beagle Juno, a growing number of programs pair animals with children. Animals visit schools, hospitals, and mental health institutions. Even over one hundred fifty years ago, the famous nurse Florence Nightingale recognized how animals could provide support for individuals living with chronic health conditions.[11]

For children who have been sexually abused, the incorporation of happy, healthy dogs in psychotherapy can lead to improvements in anxiety, depression, anger, and posttraumatic stress.[12] Teenagers who have suffered childhood traumas develop better social skills, stronger leadership abilities, and fewer attention problems when their therapy includes connecting with companion animals.[13] Hyperactivity is also quelled. Abused children who later form bonds with animals score higher on measures of empathy and self-esteem. Evidenced by physical and psychological parameters, they are often calmer, potentially interrupting a cycle of violence, perhaps with the help of the moral molecule oxytocin.

Gradually, through a combination of their own volition and legal mandates, more shelters are taking in people with animals. After Hurricane Katrina, when some people refused to leave their homes because of concern for their animals, Congress passed the Pets Evacuation and Transportation Standards (PETS) Act, providing temporary shelter for companion animals and their families during a disaster. The Pet and Women Safety (PAWS) Act will provide similar relief for abused women trying to leave their domestic partners. In addition, Sheltering Animals and Families Together, a program founded by Allie Phillips, provides resources for shelters and victims seeking shelter. Groups like Pets of the Homeless help with food, emergency veterinary care, and identifying shelters that will take in homeless individuals and their companion animals. These efforts, like Penny's, recognize the importance of the human-animal bond, in both love and its ally justice.

THE ESSENCE OF JUSTICE

There is growing evidence that justice is a biological need designed to encourage cooperation and sustain communities of people and animals. As evolutionary biologist Marc Bekoff and ethicist Jessica Pierce have pointed out, even in children and animals we see behavioral signs of justice, in the form of cooperation and reciprocity, punishment and reconciliation, and sharing and forgiveness.[14] Like adults, children and animals have negative reactions to injustice in the form of outrage and positive reactions to justice in the form of trust.

Children grasp basic rules of fairness from a young age, and their egalitarian instincts also emerge quite early. For example, a University of Chicago study showed how toddlers have different brain-wave patterns when they witness helpful, sharing animated behavior, compared with antisocial (e.g., hitting or shoving) animated behavior. They demonstrate a preference for cartoon characters who behave fairly.[15] Research has also shown how children face exclusion if they don't play by the rules. They behave according to a common set of expectations, much like the rules and patterns of behavior that guide our daily lives as adults.

Animals also learn rules of conduct. For example, primates such as chimpanzees and monkeys exchange favors and resolve conflict through measures of accountability and punishment. They console each other after conflict. One study at a zoo in the United Kingdom showed that chimpanzees would often groom or play with the most-injured chimpanzees after fights.[16] These chimpanzees were more likely to offer comfort if the hurt chimpanzees hadn't reconciled with their aggressors. Nonhuman primates also share food, and they become upset when expectations about fairness—such as an equitable food distribution in exchange for certain tasks—are violated.[17]

Dogs and their closest kin also show concern for just behavior when engaging in play. As Bekoff has written, there are four basic aspects of fair play in animals: "Ask first, be honest, follow the rules, and admit you're wrong."[18] He and his students have observed many hours of play among dogs, coyotes, and wolves. A very specific play bow, wag, and bark is often an "honest ask" to play, and animals stay closely attuned to unwritten rules during play. Larger animals level the playing field with smaller animals by softening the force of their bite and rolling over on their back. When animals make mistakes, they frequently apologize with submissive behaviors. Animals are typically held accountable if they don't play fairly, usually through scolding or shunning behaviors. I've witnessed these types of interactions among my own dogs and wolfpup, who found a home with us after she was rescued from an illegal breeder. She quickly learned the rules of fair play from our other canines, and she appeared to understand that as she grew, she needed to temper the intensity of her nips and tackles. Fights among our dogs, which are rare, tend to end in apologies and self-imposed time-outs.

The more scientists look for evidence of justice within the animal kingdom, the more they find. A reverence for justice appears to extend well beyond humans, nonhuman primates, and canines. For example, ravens intervene as third parties when norms are violated.[19] They, like many other birds, are problem-solvers. They use logic and a complex series of actions to work out problems. For instance, they can create and follow a specific sequence of steps to obtain a treat suspended on a string tied to a perch. Like we do, they test actions in their mind—an ability that likely matures with age and brain development. They can examine a problem, use their imagination, and then resolve it without trial and error.[20] These birds, like other animals, use a similar approach to resolve social problems like unfair interactions.

Since a sense of fairness appears throughout the animal kingdom, justice could serve an evolutionary or therapeutic purpose, or both.

THE JUSTICE POTION

Is justice therapeutic—does it help individuals heal? Though much of my medical work centers on securing justice for survivors of sexual assault, torture, and other forms of violence, it took me some time to consider this question.

Criminology professor Maarten Kunst and his colleagues had been asking the same question. They were interested in the idea of therapeutic jurisprudence—that legal rules and procedures can positively or negatively influence the mental health of survivors. In order to better understand this phenomenon, they undertook a large review of published studies on the subject. They asked how victims' experiences and satisfaction with the justice system influenced their emotional and cognitive well-being.[21] Despite a widely accepted notion that the justice system impacts recovery, no one had published a systematic review of empirical studies addressing the issue.

Their investigation produced a nuanced answer. Victims' satisfaction with the justice system can affect their well-being, mostly through positive changes in their thought processes. But it's not so simple. Survivors are often re-victimized by the justice system, particularly when an inadequate law enforcement response, a biased public response, or an unfair court process violates their rights. Laws, policies, and procedures can be tools of justice, but they can also exacerbate injustice.

Although study results conflict about whether the justice system is therapeutic to survivors, there is an important caveat. Regardless of the justice *system*'s therapeutic impact on survivors, the *spirit* of justice remains important to well-being and recovery. Upholding a sense of justice legitimizes suffering and offers hope to survivors. Those who seek justice often have intentions beyond healing themselves; they also aim to achieve a broader sense of accountability. Many survivors and other warriors of justice also work toward changes in power structures and reductions in abuses of power—toward upholding the principles of justice. They understand that injustice can hinder healing, whereas justice can promote healing. The reasons lie in the way the mind works.

THE BRAIN ON JUSTICE

The therapeutic value of fairness and cooperation could be explained by connections in the subcortical areas of the brain. Neurological regions also associated with love and rage, such as the amygdala, a small leaf-shaped structure called the caudate, and the ventral striatum near the base of the brain, are involved in the response to fairness. These areas of the brain receive input in the form of dopamine, a chemical that helps regulate pleasure, while communicating with the prefrontal cortex, which is partly responsible for judgment and impulse control. As a result, fairness leads to increased activity in brain territories associated with reward, pleasure, and positive reinforcement[22]—suggesting that justice is both personally advantageous and adaptive. These reactions are thought to be automatically linked rather than the result of learning—putting forward the idea that justice, in the form of fairness, is a biological need.

Teens' brains could be particularly susceptible to the presence or absence of fairness. Because of ongoing changes in brain structure and physiology, kids are incredibly vulnerable and adaptive all at once. Though the basic architecture is formed at an early age, our brains—and the brains of nonhuman animals—continue to change into adulthood. Changes in the prefrontal cortex continue into our twenties. Though children have more gray matter and therefore brain cells than they need as adults, they use fewer connections between brain cells. Around adolescence, less frequently used brain connections are "pruned."[23] A layer of fat called myelin strengthens surviving brain cells, increasing the conduction rate of messages from cell to cell. Similar changes occur in animals, though on a different time scale. Neurotransmitters like dopamine also become very powerful in the adolescent brain. As a result, habits and behaviors learned during adolescence begin to solidify—as young brains are pruned and rebuilt—for better or worse. This phenomenon could explain why some people are drawn to social justice causes when they are young, or why Sofia overtly suffered for her siblings and the animals around her.

Even from a young age, adults and peers can influence how kids develop a strong sense of "right" and "wrong." Children quickly pick up on justice and empathy cues. In the same University of Chicago study examining toddlers' responses to animated figures, children whose parents were sensitive to justice were more likely to prefer a toy that represented the well-behaved animated figure.[24] Although almost all babies discern "good" and "bad," their preferences, behaviors, and choices depend heavily on what they've learned. This and other studies suggest that kids have the building blocks for behaving in just ways, but they are strongly influenced by the environment around them. Their brains are plastic, and perceptive to our prompts, particularly as they learn to navigate the world.

As I spent time with Penny and Amber, I thought of the many young people they have reached. They see them as fledgling Phoenixes, with endless possibilities, climbing through adolescence and sifting through the cues they receive. With time, they could follow Penny and Amber's example, whose pursuit of justice for other people—and animals—has propelled their own rise. Creating sanctuary has given them strength and courage to go on, much like the Warriors and Wolves project has done for Matt. They have become a force for social change, empowering some of the most marginalized among us—a living example of Rawls's conception of justice.

THE IMPORTANCE OF SOCIAL JUSTICE FOR ANIMALS

We will likely never fully comprehend how animals experience justice, but our treatment of them has ripple effects for whether we treat each other justly or unjustly. For ages, people have questioned how animals fit into a model of social justice. Over time, laws and policies have rightly evolved toward specific protections for children and other vulnerable populations. They are protected because of their susceptibility to exploitation—their compromised power, resources, strength, or communication abilities. But peculiarly, animals have been excluded from similar protections and exploited precisely because of their vulnerability. Animals bear the burdens of human society without the benefits; a decidedly unfair equation according to established notions of justice.

As legal scholar Ani Satz has argued, existing animal welfare laws fail to protect animals adequately.[25] They are simply unjust. Though animals suffer like we do, their interests and suffering are treated differently than ours. Since humans profit from the exploitation of animals, there is a disconnect between how they are and should be treated—what Satz describes as "legal gerrymandering for human interest."[26] Even animals within the same species can be treated differently under the law. For example, my dogs are subject to more legal protection from suffering than a dog in a laboratory, begging the question of how their rights differ if their fundamental needs are the same.

Elsewhere, I have written about why I believe justice for animals is one of the crucial social movements of our time. Animals are due justice, and our treatment of them has severe consequences for the most defenseless human beings, like Sofia. How we treat animals is in many ways a marker for whether we, when privileged, will choose to intervene when the most pervasive forms of inequity are before us. Fairness requires that equal situations be treated equally. Like us, animals' experiences matter to their well-being. Since animals have complex needs and are susceptible to suffering if their needs aren't met, justice demands that we treat their fundamental needs as we would the same needs in humans—with full and equal consideration.[27] Treating animals otherwise, simply because of their easy availability and convenience, undermines justice—the instrument by which all of us, in our most vulnerable forms, can be sheltered from malevolence.

At the conclusion of a march from Selma to Montgomery, Alabama, in 1965, Martin Luther King Jr. asked, "How long will justice be crucified, and truth bear it?" He answered, "Not long, because the arc of the moral universe is long, but it bends towards justice."[28] But the arc doesn't bend on its own. It lives within us, and it bends with us. It bends how we bend. And it will require those of us unaffected by injustice to be as outraged as those who are affected.

7 : HOPE AND OPPORTUNITY

RISING WOMEN, GIRLS, AND GORILLAS IN CONGO

Hope is being able to see that there is light despite all of the darkness.

—DESMOND TUTU

As a medical consultant for Physicians for Human Rights, a non-governmental organization that shared the 1997 Nobel Peace Prize, I work with other medical and law enforcement professionals to help Congolese first responders document forensic evidence of sexual violence—just as we do in Kenya, where I first learned about Aiyana. Together, our aim is to end impunity for sexual violence in areas of conflict and unrest and help survivors secure justice. We've made progress, but there is much work still to be done.

According to the United Nations, the Democratic Republic of Congo is considered one of the most dangerous places on earth for women and girls. One woman is raped almost every minute in Congo,[1] and rape has spread amid a climate of impunity. Plagued by decades of conflict, mass rape has also become a major problem, affecting victims across the life span from infants to the elderly.

Congo is one of the largest countries in Africa, and it abounds with natural resources like oil, minerals, and diamonds. Over the past few decades, exploitation of its natural wealth has contributed to ongoing conflict and a series of escalating wars. The clashes date back to at least 1994, when Hutu *génocidaires* fled Rwanda to what was then Zaire. Thereafter, Rwanda led an invasion into the country, effectively ending more than thirty years of dictatorship and igniting a civil war in the newly named Congo. Though its civil war officially ended in 2003, the battles continue, particularly in the two eastern provinces of North Kivu and South Kivu.

Rates of sexual violence have historically been highest in the eastern part of Congo, where there is a concentration of prized minerals.[2] Though men and boys are also targets of violence, being born female has become a health risk. Congolese women have very few opportunities, expanding the silence of their voices. When their voices are raised and heard, the courage required to speak out makes them even more poignant and powerful.

During one trip to Congo, I met a woman named Neema Namadamu. She and I were asked to talk to Congolese health professionals, soldiers, and military judges about different forms of vulnerability. The group was assembled to strengthen the network of professionals responding to the epidemic of sexual violence. We met on the second floor of a large, airy, open room of an old building. Outside were fragmented roads and a string of buildings with broken red bricks.

Neema arrived in the middle of the day, wearing a dress called a *pagne* with bold patterns in brown and yellow. She easily stood out. Her husband, Danny, assisted her up the stairs of the building where the meeting was held, and once she was settled, placed her crutches by her side. Neema took the conversation where I couldn't. Though a chair was placed in the middle of the room for her, she rose up to her full height, supported by her crutches, and spoke passionately about the undiscovered power of women and individuals with disabilities. In a room of mostly stoic men in camouflage battle-dress military uniforms, with guns at their sides, Neema described how patriarchy fuels vulnerability to violence in Congo. The men listened closely to her talk on the history of conflict and oppression in Congo. With her clear message, along with her easy charm and humor, Neema drew a rousing response.

Neema's life began in a remote tribal village in Eastern Congo. As a young child, she contracted polio, causing permanent partial paralysis. After discovering Neema had difficulty walking at the age of two, her father rejected her mother and, as a polygamist, married another woman. But her mother didn't view Neema as a child with limitations. Seeing her instead as a child born with a purpose, she believed Neema should have the same opportunities her brothers and sister had. Though her mother never received any formal education, and didn't know how to write her own name, she wanted all of her children to have opportunities she did not. To pay for Neema's education, she worked long days in plantation fields.

Her drive and determination paid off immeasurably. Like all of her siblings, Neema completed high school and went on to college. She became the first woman with a disability from her ethnic group to earn a university degree—from a province where less than ten percent of women have finished primary school, even today.

When I talked with Neema about her mother, who died just a few years before as a result of untreated cancer, her electric demeanor quieted. She told me she believed her mother, like many other women, had a natural and loving wisdom passed down from her own mother.

In her book *From Outrage to Courage,* Global Fund for Women's founder Anne Firth Murray chronicles how the courage "to shine a light in the darkness and take action" makes change possible.[3] Love, she says, can be a force for social revolution and justice. Murray's sentiments intentionally move beyond the power of successful health and economic policies and into the heart of principles that guide recovery and prosperity. As she shows in her book, disrespect for freedom and personal sovereignty can lead to severe illness and even death. But Murray also highlights the importance of hope. By diving into the details of ordinary yet extraordinary lives, she shows how marginalized women can solve some of society's most insurmountable problems if given a chance. Neema, her mother, and her mother's mother all knew what Murray suggests: provided the opportunity, women can slowly change the world.

SHINING A LIGHT IN THE DARKNESS

Knowing Neema as I do, it's difficult to imagine she's ever faced doubt or discrimination. She is strong willed though tolerant, confident yet modest, patient but resolute. After graduating from college, she went on to work for the rights of women and people with disabilities. She hosted a weekly radio program and served in parliament and as a chief advisor to Congo's Minister of Gender and Family.

Neema quickly became frustrated with the system. She wanted to expand her impact—to sound her concerns about Congo to the rest of the world. She began blogging on World Pulse, a social network connecting people from around the globe. It has one mission—to "create a world where all women thrive. . . . one click, one comment, one connection at a time"[4]—and it provided a megaphone for Neema's message. But she didn't want to be the only voice. She opened a media-training center for women and girls, who quickly changed the narratives penned about them. They were no longer victims but survivors. Not weak but strong. Not just vulnerable but resilient. No longer silent but outspoken. As grassroots journalists, they presented a different version of Congo—one defined by their lived experiences rather than news headlines, shortcut slogans, or one-dimensional appeals. They named themselves the "Maman Shujaa," Swahili for "Hero Women of Congo."

Around the same time, Neema and the Maman Shujaa began working with the Global Network of Women Peacebuilders, as part of a coalition of women's groups and civil society organizations. Aiming to bridge the gap between international policy discussions and grassroots action, Neema and other peace-builders help implement key United Nations Security Council resolutions focused on women, peace, and security. These resolutions urge member nations to increase the participation of women in all efforts, and to protect women and girls from sexual and gender-based violence in regions affected by armed conflict.

Tasked with first reaching community leaders and members of local governments, Neema began conducting workshops about United Nations resolutions on the rights of women and girls. However, when Neema looked around the room of convened participants, she saw no women. Men held the positions of power. Neema faced a major dilemma: How could she reach women who didn't have a seat at the table? Many of the women she needed to reach were illiterate. And how could she fully engage men living in the same communities?

Back at the media center, some of the teenage girls began asking why Neema didn't take them to her workshops. They wanted to help. Recognizing the need for future female leaders, including the next generation of the Maman Shujaa, Neema launched the Girl Ambassadors for Peace Program, which provides mentorship for girls and young women. Soon, she had an idea to bridge youth mentorship and community education: Why not use theater to educate literate and illiterate communities on the rights and needs of women? The Girl Ambassadors were enthusiastically recruited as actors for the plays. Many people attended the productions, including village chiefs, heads of military, religious leaders, and families within each community.

In a country where almost one-third of men told one study's investigators women want to be raped and may even enjoy it,[5] attitudes have slowly shifted. Men have begun to understand the power of education and opportunity for girls and women, including health and economic benefits for their families and communities. Men in positions of power admitted to the Maman Shujaa that they hadn't realized the realm of possibilities for women and girls.

But Neema does—she continues to seek opportunities to shine a light in the darkness. In Itombwe Plateau, a very remote range of mountains in South Kivu, many girls stop attending school when their menstrual periods start, as early as twelve years of age. There, in one of the poorest areas of the world, eighty-five percent of the adolescent girls have children; many have multiple children. Without access to sanitary napkins, menstruating girls are easily identified as potential wives for forced marriages with older men, and at greater risk for stigmatization, infection, and death.[6]

After raising enough money, the Maman Shujaa sent the Girl Ambassadors to Uganda to learn how to create reusable sanitary kits. Today, they run a program out of their newly built Itombwe Center, which configures and distributes the kits along with soap made at the facility. Orders for the hygiene kits come in by the thousands, from girls who might otherwise leave school, and from grown women. With the reusable kits, the girls' attitudes have shifted from shame and stigma to pride. They now have more choice about marriage and motherhood. Gradually, their levels of confidence improve, opportunity blooms, and Phoenixes emerge. Neema told me, "There are more stories like this, and others are coming."

One of the newest initiatives focuses on children who have been shunned in their communities because of a disability, albinism, or being orphaned or born a child of rape. In one of the most dangerous parts of Eastern Congo, the Maman Shujaa have ensured that these children have the same access to education and opportunity as their counterparts. Like so many other children around the world, they seem to know education is a way out of despair—for themselves and for their country.

All the while, in the background of the same communities, Neema continues to fight discrimination, providing examples of love, tolerance, and justice. For many of the women and girls, she embodies hope. As Neema said to me, "Slowly by slowly, sanctuary is building."

But is it possible for these women to reshape Congo, changing it from the "rape capital of the world," as it is called, to a Phoenix Zone spread across the whole country?

A VISION, SISTER TO SISTER

As you might imagine, Neema isn't satisfied with the slow build of sanctuary. She never fails to realize the urgency of the situation in her homeland. While traveling in the United States in 2012, Neema was warned by friends not to return to Congo. War had broken out. On November 20, 2012, M23 rebels seized Goma, a major city in the eastern part of Congo, escalating the conflict that had ravaged the region for more than fifteen years.[7]

M23 is an abbreviation for the March 23 Movement; the name is taken from the date of the peace accords signed between the National Congress for the Defense of the People and the Congolese government. It is also known as the Congolese Revolution Army, a rebel militia group in Eastern Congo. With alleged ties to the Rwandan leadership and its military, M23 is made up of fighters who deserted the Congolese army. It was once led by a man facing war crimes charges at the International Criminal Court.

Neema was undeterred by the warnings. As my husband, Nik, once said to me as we left her home, she is a true patriot. She returned to Congo and, with the Maman Shujaa, launched an online petition one week after M23 invaded Goma. She began the appeal: "I was born in a very remote village in South Kivu Province, Eastern Congo. I belong to a marginalized tribe and I am crippled from polio. But none of those things characterize me. I have a vision for my country—a new and peaceful Congo—that compels me, and its destiny is driving me."[8]

The petition called on the Maman Shujaa's sisters around the globe—including Hillary Clinton, Valerie Jarrett, Michelle Obama, and Susan Rice—to take immediate action in solidarity with the women of Congo. In her appeal, Neema wrote about the war ravaging her homeland, her daughter being indiscriminately beaten, and other women and girls raped by governmental soldiers and militias like M23. She emphasized the need for inclusive solutions rooted in family and community—like the media center with hundreds of women activists reporting about life in war-torn Congo through World Pulse.

In the petition, the Maman Shujaa asked for the immediate appointment of a special presidential envoy from the United States to work with the African Union and United Nations to forge a peace process. They insisted that Congolese women have a seat at the table in future negotiations. The petition gradually gained traction—growing from just a few names to hundreds to thousands of signatures, finally reaching at least one hundred thousand signatories, including global leaders.

After the petition met the one hundred thousand mark, Neema received a phone call. She had an appointment at the White House the following day. There was no way Neema could get to the United States from Congo with that little time, so she called a friend. The next day, Neema and the Maman Shujaa gathered around a phone in the eastern city of Bukavu awaiting a call from Washington, where a young Congolese refugee and nursing student named Adele Kibasumba, actress and activist Robin Wright, and three others met with the National Security Council. They delivered the petition, urging the appointment of a special presidential envoy for peace in Congo. Though Neema was excited by the progress, she said, "Great. So what's next? Where's the change?"

Soon after, a US Special Envoy, former senator Russell Feingold, was appointed to the Great Lakes Region of Africa and the Democratic Republic of Congo. About one year after M23 invaded Goma, in North Kivu, the rebel group agreed to disarm. The intervention is believed to have stopped M23 from moving southward toward South Kivu, where Neema's village is located.[9]

FIGHTING THE WAR ON ANIMALS IN CONGO

In their petition, Neema and the Maman Shujaa also asked for an end to the ongoing destruction of the natural world in Congo. Home to the second largest rain forest in the world, Congo covers more than half of Africa's forestland. But the forest where Neema grew up has receded, largely due to slash-and-burn practices of cattle ranchers and the use of wood for cooking and electricity. Generations of animals who once lived there are now gone. Only domesticated animals, brought in or bred by humans, now live in the area near her village. Droughts are far more common than before, and changes in the climate are felt in real time. As a child, Neema drank the rainwater. Today the spoils of mineral mining and oil drilling taint the rain. After years of destruction, resources are scarce, deepening the burden women bear to care for their families.

Today, the Maman Shujaa work within each community to identify alternative sources of energy while leading a tree-planting movement toward reforestation. Neema hopes to see a day when animals return to their homes, which have been equally affected by conflict. In essence, she and the Maman Shujaa are trying to build a Phoenix Zone for people *and* animals living in the region.

But the Maman Shujaa aren't alone. Elsewhere in Congo, people are literally dying to protect animals threatened by the conflict in Congo. In Africa's first national park, Virunga, park rangers risk their lives daily for animals under siege there. Established in 1925, Virunga National Park is home to more bird, mammal, and reptile species than any other protected area in Africa. It has been called a microcosm of virtually all the ecosystems found on the continent. Its diverse habitats range from tropical rain forests and alpine zones to active and inactive volcanoes to savannas, marshland, and glaciers. Like many parts of Congo, it has been caught in years of armed conflict. Militias and military operations take refuge in the park, causing more havoc. Poachers have ravaged the park for the illegal international wildlife trade. Since 1994, nearly one hundred fifty rangers have been killed while protecting animals within the park. Nonetheless, few rangers quit. Though the chance of dying in the line of duty exceeds that of many other soldiers in combat, most rangers remain staunchly loyal, often reflecting the commitment of generations of dedicated ranger families.[10]

The Virunga rangers are protecting some of the world's last mountain gorillas. In 2008, a *National Geographic* cover highlighted their plight with the words "Who Murdered the Virunga Gorillas?"[11] Inside was a photo taken by Brent Stirton, a South African photojournalist. A five-hundred-pound gorilla named Senkwekwe Rugendo lay dead on a makeshift open casket carried on the shoulders of rangers and villagers. Like other members of his family, he had been shot and killed in the summer of 2007, at a time when paramilitary groups, rebel troops, and the Congolese army occupied the park. Some of the rangers knew the gorillas as well as their own families. Like the trackers in the Kibale forest in Uganda—one of the sites where primatologist Debra Durham and I conducted our study of mental disorders in chimpanzees—the Virunga rangers spent every day with the gorillas, getting to know their distinct personalities.

Sometime during the night, the killers, armed with automatic weapons, hunted down Senkwekwe's family. The gorillas had become so habituated to humans that they were likely shocked by the betrayal. By the next morning rangers had found three female gorillas—Mburanumwe, with her unborn baby; Neza; and Safari—all shot to death. Safari's body was burned by fire, and her infant crouched nearby. The following day Senkwekwe, the patriarch of the family, was found dead. He had been shot in the chest. Three weeks later the body of another Rugendo family female, Macibiri, was discovered. Her infant, Ntaribi, was too young to survive alone.

That wasn't the first attack on gorillas in Virunga. One month earlier, two females and an infant from another gorilla family were attacked. One of the female gorillas had been shot in the back of her head. Her infant was found clinging to her breast.[12]

Eventually, authorities identified those responsible for the massacre. The shootings were linked to an illegal charcoal trade inside the park.[13] The gorilla murders were meant to signal a warning to stop investigations relating to the trade. Some of the rangers were likely involved in the killings, and they were prosecuted or fired. Congolese authorities arrested the director of the park in connection with the slaughter.

Since 2008, the new director, Belgian prince Emmanuel de Merode, has helped set a new vision for the park, its rangers, and its animals.[14] Though armed poachers and members of militia outnumber the park rangers ten to one, their work is slowly paying off. The park's mountain gorilla population appears to be increasing. And the number of rangers is growing, too—almost doubling in numbers. For the first time in history, some of the new recruits are women, and they are considered equal to the men.[15] They are also as committed as the men. After her colleague was killed, ranger Salange Kahambu told CNN, "We always say in life, those who risk nothing, have nothing."[16]

In an interesting twist of events, Virunga has become a force for peace and hope in Congo and the surrounding region. The Virunga Alliance has negotiated alternatives to using resources in the park, helping increase economic and other opportunities for people living around the park. By creating hydroelectric plants outside the park and other small pilot projects, they intend to create one hundred thousand sustainable jobs in the area within a decade.[17] They also aim to help the four million people living around the park improve their access to drinking water, education, and health services. The Alliance offers a bright light in a region of the world rocked by conflict and desperation, much like Neema and the Maman Shujaa do. But they offer more than just hope. They offer a path toward healing.

HOPE, RESILIENCE, AND THE PLACEBO EFFECT

In *The Anatomy of Hope,* Jerome Groopman writes about his thirty years as a physician treating patients with cancer and other life-threatening diseases. He describes how hope inspires "the courage to overcome fear, and [solidify] resilience."[18] Throughout his book, he explores its vital role in people's lives and recoveries—much like what it represented for Grace and Doc.

Hope and optimism—a belief in and expectation of "the best of all possible worlds"—can protect against physical and mental illness.[19] Over time, researchers have developed various tests and scales to measure hope and optimism.[20] For example, an optimistic bias, a sign of hope, signals that expectations are better than reality, whereas a pessimistic bias indicates that expectations are worse than reality. Whether measured as a constant state or temporary trait, hope is linked with positive outcomes among individuals of myriad backgrounds and experiences. People with more optimistic dispositions are less likely to be depressed or suicidal, whereas people with pessimistic outlooks are at greater risk for depressive and anxiety disorders and a poorer quality of life.

After trauma, hope can reduce symptoms of PTSD. By decreasing stress, it can protect against fear and foster better coping strategies. It also protects against physical problems. Optimistic individuals experience more rapid recovery after short-term medical illnesses and better survival rates after heart attacks and cancer diagnoses, compared with their pessimistic counterparts. There is even evidence that hope can help communities recover after conflict.

At the suggestion of a colleague, Jerome Groopman began his journey to understand the healing powers of hope by investigating the placebo effect, which also operates on belief and expectation. The placebo effect, a surrogate for hope, can actually mimic the effects of morphine. It can calm us and block pain by causing the release of endorphins and enkephalins, naturally made within our brains.[21] Positive beliefs and expectations can set off a chain reaction by diminishing pain and anxiety, allowing hope to expand, further lessen pain, and speed up recovery. But optimism isn't born or nourished in a vacuum. It is often determined by how free we are to make decisions about our own lives and futures. As Groopman points out, hopelessness and pessimism can trigger the opposite chain reaction—as decades of experiments inducing helplessness and hopelessness in animals in laboratories have also shown. Together, hope and freedom help form the fulcrum of the pendulum between vulnerability and resilience.

As with other emotions that fuel healing, hope can also cause the release of dopamine, which works with other chemicals deep within the reward centers in the brain.[22] Multiple regions of the brain must work together to process factors that influence beliefs and expectations, including brain structures that help sort through fears, store memories, and regulate stress. Through complex interactions, it's possible that hope sparks new brain connections and reawakens dormant areas of the brain. But first, there must be opportunity—like what Neema and the Maman Shujaa have created for girls and women in Congo, and what the Alliance has provided for thousands of people and animals living around Virunga National Park.

But is it possible that hope and opportunity also matter to animals living around Neema's village and the gorillas within Virunga? If so, what might we learn?

HOPE IN ANIMALS

After reading Groopman's book, I remembered a story about a chimpanzee named Bobby. Before he was released to a sanctuary, Bobby was used in laboratory experiments much like Negra was. He was injected with diseases, forced to undergo repeated biopsies, and kept in an isolation chamber for prolonged periods of time. At the sanctuary, he was given anti-anxiety drugs to treat his compulsive behaviors and severe self-mutilation. After each dose, he calmed down. But, concerned about the effects of too many anti-anxiety drugs, his caregivers decided to intersperse some of his doses with saline—a placebo. Something surprising happened. After receiving the placebo, Bobby calmed down, just like he did when he got the real thing. At the time, I thought there could be two explanations for his behavior: either he was acting, tricking those around him, which chimpanzees have been known to do; or he experienced the placebo effect.

As it turns out, the placebo effect has been described at length in animals. In a 1999 article in the *Journal of the American Veterinary Medical Association*, veterinarian and behavioral expert Frank McMillan outlined three main mechanisms that could explain how the placebo effect manifests in animals.[23] The first is via the conditioning effects of placebos, which we also experience. Experimental psychologist Ivan Pavlov was the first to report on a conditioned placebo effect in animals. He gave dogs morphine in an experimental chamber and recorded their specific individual responses. Afterward, just placing them in the experimental chamber produced the same reactions. The experimental chamber acted as the placebo. Today, a visit to the veterinarian's office might produce a similar effect, whether it's a positive or negative association. Since Pavlov's early experiments, a conditioned placebo effect has also been noted in monkeys, dogs, guinea pigs, rats, and other animals using narcotics, insulin, and anti-epileptic and immunosuppressant drugs. These findings imply that animals experience a mind-body connection like we do.

Animals also form beliefs and expectations—the two components of hope seen in humans. Creating expectations depends on conditioning, learning, and related cognitive mechanisms—all qualities animals possess. It's possible that optimism in animals is explained by brain mechanisms that are very similar to those found in our brains. For example, the "chain-reaction, hope-inducing" endogenous opiates found in our brains are also found in the brains of other animals. Like us, animals can develop an optimistic or pessimistic bias.[24] Their collective experiences, memories, and reflections on their memories can influence whether they become pessimistic or optimistic, and how they make decisions. They learn what to expect about the future, and expecting relief from discomfort can provide a sense of control and more hope. Positive experiences and expectations can alleviate helplessness and hopelessness and lead to better perceptions and choices, as Scott Blais has observed in elephants freed from chains. Sometimes referred to as a cognitive or judgment bias, optimism and pessimism are increasingly being used as markers of well-being in birds and mammals.[25]

One day I was talking with my husband, Nik, about whether animals experience hope. He said something rather intuitive: "What about all the animals who get adopted from shelters—like ours? The looks on their faces and the gradual changes in their demeanor say it all. Their beliefs and expectations change." As I thought more about the despair animals often experience, and how there must be two sides of a coin—in this case, despair and hope—his comments made more and more sense. Nonetheless, like us, to be optimistic about the future, animals must also have the right opportunities.

But what does opportunity mean for animals? And what does it really mean for us?

For the gorillas now living at Virunga, opportunity is the opening of a door to a species-typical life—one that is free, sovereign, and filled with loving social bonds and the chance to live up to their potential. When given the chance to live a full life typical of their species, animals are healthier and more resilient. They are more likely to thrive mentally and physically. The same is true for us. Though our precise needs as individuals and as a species differ, the principle is the same.

Imagine a world that allows for the full self-determined potential of every being. Consider what we might learn from each other—as women and men, girls and boys, nonhuman and human animals—and how we could grow together. Envision a Phoenix Zone, truly planetary in scale, replacing what were once conflict zones.

8 : DIGNITY

SAFE HARBOR FOR DEGRADED PEOPLE AND FARM ANIMALS

It seems there is no respectable way to deny the equal dignity of creatures across species.
— MARTHA NUSSBAUM, *Frontiers of Justice: Disability, Nationality, Species Membership*

Though Congo is a conflict zone, it isn't the most dangerous place on earth for animals. In the United States, spread across almost every state, large windowless warehouses have become the deadliest conflict zones for animals.

Beneath the surface of every conflict zone is a mix of subjugation, exploitation, and victimization. As fights for domination and power ensue, individuals and whole societies become scarred by mass atrocities. But where did our propensity for mass crimes like rape and genocide arise? What makes us capable of erasing the identities of whole groups of people? Of standing by while others perish, ignoring their suffering, their dignity—their value? Can we trace the origins of mass violence, like what has happened in Rwanda and Congo, to unravel its power?

Is it possible that some answers lie in our treatment of animals?

In his book *Eternal Treblinka,* Holocaust historian Charles Patterson compassionately shows how the domestication and exploitation of animals became the model and inspiration for many forms of human oppression that followed.[1] Before writing *Eternal Treblinka,* Patterson became close friends with a Jewish refugee who had fled Nazi Germany. That spurred him to study the Holocaust, which in turn led to his first book, *Anti-Semitism: The Road to the Holocaust and Beyond.* Later, after learning about the plight of animals in society, he wanted to understand the shared history underlying the maltreatment of people and animals.

Contrary to more controversial comparisons between human and animal suffering, Patterson doesn't necessarily suggest that our treatment of animals resembles the worst atrocities against humans. Rather, he argues, some of the worst atrocities against humans resemble our treatment of animals. Therefore, we need to unravel the unjust treatment of animals in order to understand and prevent inhumane acts against humans. For example, the denigration and subsequent torture and murder of Holocaust victims, Patterson argues, closely followed patterns seen in the American slaughterhouse and Henry Ford's slaughterhouse-inspired assembly line, abutted by Ford's racist propaganda.[2] Though some might be resistant to Patterson's views, due to concerns about comparing the experiences of people and animals, other historians, anthropologists, and legal scholars—along with Holocaust survivors—support his conclusions.[3] In the last half of his book, Patterson shares stories of survivors and their children who became advocates for animals.

Even more than a century after Upton Sinclair published *The Jungle,* based on his investigation of the Chicago meatpacking industry, there has been little meaningful change in the ways farm animals experience life and death. On the contrary, animals' lives and deaths have become more mechanized, removed from human view and emotion. Since the beginning of the twentieth century, human beings have also faced two world wars, the rise of genocide, and escalations in the use of rape as a weapon of war. Patterson claims these phenomena are not unrelated.[4] Nowhere is this link more visibly apparent than in industrial farms and slaughterhouses—what Sinclair described as "a dungeon, all unseen and unheeded, buried out of sight and of memory."[5]

I was reminded of the connections between the treatment of people and animals during a conversation with Susie Coston, director of Farm Sanctuary, the largest sanctuary for farm animals in the United States. Over vegan wine, nut cheese, and crackers, Susie described a haunting audiotape recording she first listened to between flights from California to her home in New York. The tape was forwarded to her after a worker at a pig farm provided the tape to local law enforcement. As Susie listened to the recording, she heard what sounded like a woman being attacked by a group of men. Susie cringed at the derogatory language, blows, and screams echoing off metal walls. As she soon learned, the crying pig—later named Julia—was beaten with a metal rod and burned with an electric shock prod. The assault ended only when Julia collapsed to the ground. The beating served as her punishment for refusing to move from an iron gestation crate, a cage where pigs live after they are forcibly inseminated, to a farrowing crate where she would be forced to labor, nurse, and relinquish her babies—a trauma she had lived through at least three times before. After falling to the ground, Julia was dragged into the farrowing crate. Like in the gestation crate, there was barely enough room for her to stand or lie down between the bars of the small cage. Julia was only two years old at the time.

At the request of the local police department and the local Society for the Prevention of Cruelty to Animals, Susie pulled together an emergency rescue team for Julia. Like other industrial pig farms, the building was made of metal and concrete and devoid of sunshine, mud, or grass. As Susie entered the building, pigs trapped in small gestation and farrowing crates—just like Julia was—surrounded her. Though her job was to rescue Julia, Susie kissed every pig she could on the nose. Like Julia, they mattered. Her story differed only in that she made it out.

ERASING IDENTITY

Julia was in a farrowing crate, though she had not yet given birth, when Susie and her team arrived at the industrial farm. They immediately took her to Cornell University for veterinary care. On the way, Julia appeared exhausted and terrified. Burns and lacerations covered her head, back, and ears. Bruises covered her belly. She slowly slipped into shock. Her story reminded me of my ordeal with Love, the dog I met halfway across the world in Kenya. As Susie told me Julia's story, I wondered if the Cornell veterinarians documented her wounds as forensic evidence for law enforcement, as I do for human torture survivors—proof of the electric prod drawn up her back, the sting of a whip against her body, or the repeated kicks to her side as she lay on the ground. I doubted her flat feet and leg sores from the crate bars and coarse concrete floors would be recorded as evidence of abuse, since the conditions she was kept in were considered legal, descended from a system of domination that began in the Middle East over ten thousand years ago.

Julia is one of many. An estimated sixty billion land animals are killed for food each year. Nine to ten billion animals live and die each year in the United States, where factory farms are ubiquitously invisible from New York to California. Many more sea animals are killed each year in farms and the ocean. Driven by a legal mandate by the US Secretary of Agriculture and government subsidies, factory farms—from small, family-owned businesses to giant corporations—now hold more than ninety-nine percent of mammals and birds raised and slaughtered for food.[6] They are packed so tightly in cages that they can't walk, turn around, or stretch—let alone escape, withdraw, or reach for another being. As their minds are flooded with fear and pain, their bodies are placed on conveyor belts, suspended from hooks and chains, and slit open as their personalities, their identities, disappear into the stench of death on a mass scale.

But they aren't merely numbers. Like people and chimpanzees, farm animals can have rich lives. They can also experience PTSD, depression, anxiety, and compulsive disorders. They are someone, despite attempts to mold them into an image of no one, en masse.

As Patterson describes in his book, before domestication, human societies commonly held a deep kinship with other people and animals living around them.[7] However, domestication brought "detachment, rationalization, denial and euphemism"—words echoed by Alfred McCoy in *Torture and Impunity* in his description of enhanced interrogation techniques against prisoners.[8]

The exploitation of animals for their meat, milk, and skin emerged relatively recently in our evolutionary past—about eleven thousand years ago in the ancient Near East.[9] Over time, humans captured and domesticated young animals, first by killing adults protecting young ones. In the process, humans developed methods to control the lives of animals—from birth to death. They used whips, chains, spears, and knives. They turned to castration, insemination, and branding to further master their lives. And still, to cropping off portions of their bodies: snouts, ears, and tails. Their status was denigrated to an object of oppression. As a result, animals lost their freedom, sovereignty, loving social bonds with their families, and fair opportunities to live a normal life—all in the name of dominion. This notion would become entrenched in law, religion, and science, and summarily articulated in Aristotle's "Great Chain of Being"—a strict hierarchical structure of all matter and life, supposedly decreed by God.[10]

Over the past several millennia, the exploitive treatment of animals spread across the world, particularly in the form of industrial farming and slaughterhouses, while human subjugation also took root. Patterson notes, "Since violence begets violence, the enslavement of animals injected a higher level of domination and coercion into human history by creating oppressive hierarchical societies and unleashing large-scale warfare never seen before."[11] Those in power adopted the same practices used to govern animals—from castration, branding, and mutilation to whipping, collaring, and chaining—to control enslaved humans. Today, these methods remain in place where torture and modern slavery are practiced.

The hierarchy and great divide built on the domestication of animals also planted a seed that some people could be reduced to something "less than human"—"beasts," "brutes," and "savages"—an ethic that Patterson says has encouraged colonialism and genocide.[12] A refusal to acknowledge the individual lives of animals has severe consequences for humans. If we cannot value the identity of an animal, we cannot value our own identities in our most naked, vulnerable forms. It is far more difficult to empathize with those we don't find worthy—in many cases, those denigrated to the status of an animal, an "ominous sign," according to Patterson.[13] It sets people up for humiliation and exploitation, which often leads to murder and mass bloodshed—as it did for those referred to as "Rajah" (cattle) during the Armenian genocide, Jews called "Ratten" (rats) in Nazi Germany, and Tutsis labeled "Inyenzi" (cockroaches) before they were slaughtered during the Rwandan genocide. Only by treating humans as we treat animals can perpetrators commit such horrific atrocities: when people are bare, huddled together, crowded like animals, at their most vulnerable, and stripped of their personhood.[14]

In Laura Hillenbrand's *Unbroken,* the remarkable tale of Louis Silvie Zamperini, an Olympian runner and prisoner during the Second World War, she describes how prison guards sought to deprive Zamperini of dignity. To be divested of dignity, Hillenbrand writes, is "to be cleaved from, and cast below, mankind." And "[w]ithout dignity, identity is erased."[15] But the overarching message of her book suggests something else: it's the dignity and identity of those who cause and tolerate oppression that is diminished.

In Susie's fight to save those like Julia, she offers us all a new potential identity, and a more dignified existence. Rather than erasing them from our collective consciousness, she is drawing attention to their lives, their identities, their dignity.

I met Julia in the summer of 2016, four years after she was rescued. I exited onto an old country road in upstate New York, eventually turning right onto a gravel road. At the end of the road, I stopped and looked to my right, where I was reminded of friends and family who had joined my husband and me for our wedding celebration there twelve years earlier. When we lived in New York City, Farm Sanctuary became our refuge from a city of dizzying distraction. Ours was the first wedding scheduled on the premises, and watching our families and friends step carefully across mud and grass in formal attire was worth it.

As I got out of my car and looked up, I saw the same clear blue sky I recalled from over a decade before. I stepped into the brick-red "People Barn" reserved for gatherings, where we had danced into the late hours that night. I asked for Susie and was directed to the hospital where rescued animals are triaged. Susie, rising to a height of five feet four inches, and her dogs greeted me. Smiling light-green eyes framed by long brown hair disguised a sweet toughness about her. Susie was kind enough to take me on a tour of the sanctuary with another thoughtful woman who worked for the *New York Times* and with Doctors without Borders.

We met several staff members, all of whom adore Susie. For many of the staff members, the farm has become their sanctuary too. It's easy to see why. Even after working with tens of thousands of animals over fifteen years, Susie knows every name and story—from her staff of almost forty people in New York and California to the one thousand animals at the nation's three sanctuaries. Freedom, sovereignty, love, justice, and hope permeate the refuge. Chickens who once lived in battery cages walk confidently down the dirt road in the middle of the sanctuary, as the sun shines down on their faces. When Susie calls their names, they hurry over. Cows rescued from beef and dairy farms call to each other in the distance. Sheep and goats saved from filthy "organic, humane" farms nudge each other in warm barns.

I learned more about Julia's story on our tour. After receiving care at Cornell, Julia was taken to Farm Sanctuary. Immediately after arriving, she went into premature labor. She developed painful mastitis and had trouble feeding her babies. The staff poured themselves into her care and recovery. And slowly, with time, Julia recovered. Though she's had multiple medical problems and surgeries related to her time at the industrial farm, she got a chance for a life her pig sisters didn't have. She ushered her babies into the world. She's gotten to know and bond with them as they've grown up. When she was separated from her kids after her surgeries, they insisted on lying right by her gate so they could touch each other at night. Since her piglets were born at the sanctuary, they've never known the reality of a dim, crowded warehouse, docked tails, assembly lines, or the sights, smells, and sounds of mass mortality. They've all lived beyond the six months their human masters carved out for them in this world, when most pigs are killed for their meat. Instead, they have opportunities in life, liberty, and love.

Though Julia's babies are unaffected, she still struggles with memories of her trauma. Like other pigs with compulsive disorders who bite bars of their cages, she chews on rocks. As a result, her teeth are severely worn and decayed. She's more fearful than other pigs at the sanctuary, and she avoids situations that could remind her of the suffering she endured. Sometime after Julia found a new home at Farm Sanctuary, a man who worked at the industrial pig farm visited her. He was the person who first alerted authorities about Julia's assault, though he was also part of the oppressive system that kept Julia squeezed between the bars of a small cage. As he approached the sanctuary fence, Julia aggressively charged forward. He backed away and soon left the sanctuary. Though her response might have been a flashback related to PTSD, the fact that she had the liberty to assert herself—and be heard—was telling.

Today, Susie believes Julia and her babies are living "a life that every animal deserves." She knows the animals are sentient beings with their own individual worth, not just numbers or commodities. They are someone, not something.

Around the time of Julia's arrival, Farm Sanctuary launched the "Someone Project," which aims to show who animals really are. As science has shown, many animals are self-aware. Like us, animals have a sense of self. Many animals can identify themselves or others by sight in a mirror, through odors or sounds, or via an integrated composite of sensory input. Since scientific discovery always lags behind reality, it's likely that most, if not all, animals are far more self-aware than we realize. Any animal who can perceive environments and conceive of a future probably has some form of self-awareness.[16]

Using the latest science, the Someone Project sheds light on the unacknowledged and undervalued mental lives of farm animals like Julia—from their smarts to their social skills. For example, scientists have shown that sheep have complex emotions—from fear and rage to despair and joy—and that they are far more intelligent than they have been given credit for.[17] Goats have sharp senses of humor and incredible tenacity, though like all animals their individual personalities differ.[18] Chickens—who are blinded from the waste in industrial farms and crippled by reproductive cancers due to breeding practices—can anticipate their future and demonstrate self-control, advanced cognition, and abstraction.[19] And like chickens, us, and other animals, cows learn from one another. They have high social IQs, strong spatial memories, and complex problem-solving abilities. Cows like Ari, the formerly lifeless child of a discarded dairy cow found at a Pennsylvania stockyard, and his good friend Nik, a bull calf who broke loose from a farm, are guided by compassion, and they become excited when they meet someone or learn something new.[20] Today, though Nik, named after my husband, and Ari are from two different worlds, they don't leave each other's sides at Farm Sanctuary.

And Julia isn't the only intelligent, emotional pig. As independent researchers have shown, pigs use tools. They recognize names and understand that mirrors are reflections instead of windows. Playing abstract video games, they can outwit one another and solve complex problems.[21] They also develop strong bonds of friendship and family. At times, their capacities even appear to exceed our own—as shown by Rose, the first pig Susie met who had been in a gestation crate in a massive factory operation. She even changed the way Susie saw pigs.

LEARNING FROM ANIMALS

Between June and July 2008, most of the rivers in eastern Iowa flooded. Although there were no human fatalities recorded, countless numbers of animals died. Though these animals' deaths were recorded in the billions of dollars lost, little is known about their actual fate. Rose's story is different. She was found with her dead baby piglets who had died of starvation, their emaciated mother unable to produce milk. In a strange way, the flood saved her life, though her piglets weren't as fortunate. When rescuers found Rose, she was attempting to rouse her piglets by rolling them with her nose and pushing them. Rose was starving to death but still trying to wake her babies, just as a human mother might try to awaken her dead infant. Maternal pig love is fierce. Recalling the first time she met her, Susie said, "Her face showed such a deep sadness . . . I had never seen pigs so broken-down and clearly so afraid."[22] Rose's eyes were empty.

Even after arriving at the sanctuary, Rose maintained a blank stare and withdrew socially, signs of depression. However, with time, Rose met another pig at the farm named Nikki, who had piglets with her. Like Rose, Nikki was found on a levee during the floods, exhausted and wasted, protectively nursing her newborn babies. When rescuers approached her and her babies, she cried out in alarm. At Farm Sanctuary, Nikki was so protective that she tore apart gates to get to her piglets. At first she even tried to protect them from other pigs, like Rose. But Rose was obsessed with the piglets. She tried to get to them and talked to them through the fence. Eventually, Nikki allowed Rose to co-parent the piglets. Rose and Nikki developed a friendship, and Rose remained with Nikki and her piglets. At times, Rose was even more protective than Nikki. She wouldn't go to bed until all the babies were tucked in and secure. When one of Nikki's babies, Chuck, intimidated his sister Rory, Rose stood by her. Rory was afraid to come in at night, so Rose would go out and sleep beside her, never leaving her alone. Slowly, with Rory's love, Rose became more communicative and social with other pigs and humans, and her depression slowly melted away. Rose and Rory became inseparable until the day Rose died.

In many ways, stories like Rose's and Julia's might be leading to change. In the European Union, a ban on gestation crates in 2013 followed a full ban on veal crates in 2007 and a ban on battery cages for hens in 2012. In the United States, a number of states have adopted similar laws. But is this enough? Or do we owe them more than a ticket out of a gestation crate into just a larger, cleaner cage? As Susie notes, "Like Rose, all pigs have the capacity and desire for complex, loving relationships with each other. And like Nikki, all pigs have an understanding of the perspective of others and are able to put their observations into practice—in Nikki's case, understanding Rose's interest in her babies and permitting her to get close to them."[23] Sometimes I wonder what capacities Rose, Julia, and Nikki have possessed that young Sofia's mother didn't. Did they value their children more? Recognize the dignity in their very existence? What could we learn from animals like them—including their profound capacities for love, empathy, and resilience—if we valued them for who they are rather than demeaning them as less than they are?

By denying their individual worth, what damage are we doing to our own dignity? What if we were to see them more as Abraham Maslow appeared to slowly learn to see animals, through their capacities and complexity, and as Phoenixes who can rise from the ashes?

DIGNITY—FROM RANK TO INDIVIDUAL WORTH

Though the hierarchy and great divide built on the domestication and exploitation of animals fostered an idea that some humans could be reduced in worth, it's possible that recognizing the independent value—the dignity—of animals could in turn lead to a more just world for humans.

Historically, dignity was tied up with rank among people.[24] A queen did not merit the worth of a king, and a laborer did not match the consideration of a lawyer. A child's interests did not weigh as much as an adult's interests, and race, gender, and class determined one's status in society. With time, this division became enshrined in the law. People in positions of power were protected over those without authority. In many places around the world, these ideas persist. Power frequently results in a higher level of recognition, rights, and privileges in society.

Reacting to atrocities committed during the Second World War, human rights proponents began to promote the concept of inherent human dignity, as both a principle of morality and a principle of law.[25] Dignity has become the basis for legal rights, as the architecture for both affirmative rights—like economic, social, and cultural entitlements (love, opportunity, and hope)—and negative rights—like protections from abuse, exploitation, and discrimination (threats to freedom, sovereignty, and justice).

In an effort to differentiate people from animals in political, legal, and cultural contexts, human dignity regularly denotes a special elevation of the human species. Defenses for this separation often rely on the assertion that humans have specific capabilities, such as reason, language, and morality.

However, scientific discoveries increasingly show that we are not alone in our capacities. Nor are there "higher" and "lower" species, but instead individuals adapted to their own natural environments. Like us, animals have varying degrees of rationality and language. They are intelligent, have sensitive social instincts, and display a sense of morality and justice. They can be both empathic and altruistic. Nonetheless, we continue to search for proof that we are worth more than other animals—from our opposable thumbs to the way we walk and talk. This approach isn't unlike attempts to distinguish humans from each other based on genetic or anatomic differences. Nor does every human walk upright, grip with a hand, hear, or talk—whether because of differences at birth or changes in capacities over the course of a lifetime. These distinctions say nothing of one's value.

Here, there is another grave issue we cannot ignore. The social and legal elevation of humans above animals hasn't stopped mass atrocities, crimes against children, torture, or sexual violence. If we can't find something to separate us from others—to justify the injustices committed against one another—we resort to referring to other people as animals. But what if we were to value others simply based on who they are—for their moral individualism, in which one's own characteristics constitute morally valuable power?[26]

Since we, as humans, don't possess all of the capacities of animals of all species, we can never really comprehend their full, lived experiences. But we should at least honor what we know of them—that they need the freedom to determine their own lives and develop relationships with others in their own species, while living up to their potential.

As legal scholar Martha Nussbaum suggests, dignity should be rooted in the value that originates in each one of us.[27] If we are honest with ourselves, our regard for human dignity is not grounded in rationality. Instead, we find dignity in our animality, our vulnerabilities, and the ways we can each thrive in the bodies we live in, if given the chance. By respecting each other's liberty and sovereignty, as well as the need for love, justice, and opportunities to flourish, we can honor the dignity of each individual. The case is no different for animals. This is perhaps the essence of the Phoenix Effect. We aren't just cogs in a wheel—in life or death.

THE THERAPEUTIC VALUE OF RESPECT FOR DIGNITY

Respect for dignity is critical to the mental health and resilience of individuals who are sick or suffering. For example, using data from the Commonwealth Fund 2001 Health Care Quality Survey of more than six thousand adults living in the United States, researchers showed how treating patients with dignity improved well-being and specific health indicators.[28] But, perhaps somewhat poignantly, the end of life is the area of medicine that has garnered the most interest in dignity.

For the past several decades, professionals who care for dying individuals have tried to find interventions to help people cope with the reality of death. More than fifteen years ago, psychiatrist Harvey Chochinov created "dignity therapy," which aims to ease the suffering of dying people and their families.[29] When a man with an inoperable brain tumor showed him how he needed to be seen, Chochinov realized the importance of dignity to mental wellness.[30] Over time, dignity therapy has evolved to address the capabilities and priorities people each have at the end of life. For many, these concerns include respect for basic freedoms, choice, social needs and connections, and individualism. Over the past decade, researchers have studied the effects of dignity therapy in North America, Europe, Asia, and Australia. They have found that respect for the value of each individual fosters well-being, and it may lower levels of anxiety and depression, even while dying.[31]

Susie's experiences with animals start from a different place; she shows respect for their dignity by giving them life. As a result, animals at the sanctuary begin to express normal behaviors, engaging areas of their brains long suppressed, as they pursue the lives their bodies and minds were born to have. Susie envisions a world where billions of animals aren't seen as mere commodities. She knows the current treatment of farm animals has implications for how we value all individuals with inner lives, including our fellow humans.

DIGNITY AND LASTING SOCIAL CHANGE

The title and vision for Patterson's *Eternal Treblinka* stem from Isaac Bashevis Singer's short story "The Letter Writer," about a man who lost his entire family in the Holocaust and bonded with a mouse who visited him at night. In eulogizing the mouse, Singer's chief character declared,

> What do they know—all these scholars, all these philosophers, all the leaders of the world—about such as you? They have convinced themselves that man, the worst transgressor of all the species, is the crown of creation. All other creatures were created merely to provide him with food, pelts, to be tormented, exterminated. In relation to them, all people are Nazis; for the animals it is an eternal Treblinka.[32]

Like the character in his story, Singer's family members were killed during the Holocaust. By following his brother to the United States, he survived. He often wrote about Holocaust survivors and refugees as well as the plight of animals.[33]

It is somewhat striking that Singer chose to write about a mouse, an animal who has been demonized and exploited like few others in human society. Our empathy for them often appears nonexistent, despite many studies showing how profoundly empathic they and their equally demeaned kin, rats, are. Repeatedly, scientists have shown how mice and rats cringe in pain and try to save their peers when they are intentionally terrorized through severe confinement and the infliction of pain.[34] Researchers often justify these experiments by suggesting that they teach us something about who we are as humans. But what do they really say about us, our dignity, and our own capacity for empathy?

What if we removed ourselves from the crown of creation? Could we then perhaps abandon our transgressions—not only against animals but also against one another?

When I speak with my friend Neema about power, she is very clear with me. I ask her how the women and girls she works with are able to change the minds of men in positions of authority. She tells me, "Those men are not powerful; they are in the system. And if we change, the system will change around us." She leverages the subtly powerful roles women have in their communities to influence change. I believe Neema discovered something long ago, just like the author Brené Brown did: true power—that which has the force to heal the world—is found in our vulnerabilities, and it is often strengthened through struggle. Those with false power cower at the door of true power, wishing to oppress it and keep it from rising. But it will rise, if only we recognize the value of each individual—a woman once buried by oppression; a tortured refugee pursuing a better life; a child seeking justice; a loving warrior and wolf; a chimpanzee freed from her cage; a sovereign elephant no longer bound by chains; and, yes, a pig and her adopted daughter.

PART THREE

The Rise

BUILDING PHOENIX ZONES
IN A CONFLICTED WORLD

9 : WHAT CAN WE LEARN FROM PHOENIX ZONES?

A people that values its privileges above its principles soon loses both.

— DWIGHT D. EISENHOWER

My friend John Gluck, who worked years ago as a primate researcher and turned later in life to animal protection, frequently talks with me about moral residual, a lasting responsibility to ameliorate the harm we create. His reflections borrow from his interactions with Governor Robert Lewis of the Zuni Pueblo. He once told John how one winter, as a child, his father taught him to reciprocate for the pinion nuts he had taken from pack rats' nests. Through his father, Lewis learned he was responsible for reducing the vulnerability—the moral residual—he had created.[1] John often refers to this story when considering his personal moral obligations to animals, as well as society's obligations to them.

We, as humans, have the incredible potential to be both constructive and destructive. Though we are capable of astonishing invention, we are simultaneously inclined to simplify our place in the world. To forget our interdependence and responsibilities to other people and animals, seemingly immune to the laws of nature and the moral residual we leave behind. But we are not invulnerable to the vulnerability we create; it follows us in much the same way structural violence does. It must be addressed at one point or another.

The world left behind is the one we inherit. Our failures in empathy have consequences. Old problems become new, subject to manipulation or resolution. No matter what the future holds, we will never completely undo the damage that has been done. The suffering. The violence lived and felt by real human and animal beings, and the weighty residual it has imprinted upon us.

But we can move beyond our history of violence and establish a better path ahead. In the zones of the Phoenix, we are capable of survival, recovery, and expansion. Progress is not a zero-sum game. Each zone produces momentum for the next. Their unfolding, an avenue to change. And their diffusion offers space to avoid the creation of harm in the first place. But like their architects, we need to enter with open hearts, honest intentions, and humble and nimble minds. With a willingness to question our assumptions. To right our wrongs. To rise from the ashes.

Over time, the legend of the Phoenix has grown. Most forms of the myth reflect a basic narrative: after hundreds of years, when the old Phoenix feels her death approaching, she begins to collect aromatic plants to build a nest in a sacred tree.[2] In the more well-known tradition, the nest is heated by the sun, igniting the Phoenix. From the ashes, a new Phoenix arises. One of the earliest recorded representations of the Phoenix has been traced to the ancient Egyptian Bennu bird, which loosely translates to "rising brilliantly" or "shining," much like the sun. In literature, the Phoenix has come to represent an enduring sense of redemption and recovery.

The myth of the Phoenix offers a powerful example of how we can rewrite our personal and collective narratives. The Phoenix Effect is a metaphor for how we can heal ourselves as well as the world around us. We can reimagine our future. Tragedy and crisis can be reshaped into opportunity and hope. In trying to better understand how this transformation is possible, over time, I looked for patterns where some of the most wounded people and animals among us recover. Every Phoenix Zone led me to a set of pillars—comprehensive and fundamental principles. Each honored respect for basic liberties and sovereignty, a commitment to love and tolerance, the promotion of justice and opportunity, and a belief that every human and animal possesses dignity. They are where the wounded rise, and where great care is given to creating peace, beauty, and strength for those who have been hurt and humiliated. Where asylum seekers find sanctuary and chimpanzees find freedom. How enslaved people and elephants become sovereign. Refuge where combat veterans and wolves together overcome fear, PTSD, and depression. What homeless youth need to overcome histories of abuse and neglect. And where Congolese peacemakers help women and gorillas in one of the most dangerous conflict zones on earth. Phoenix Zones offer lessons in how respect for the individual value of each human and nonhuman animal—including discarded farm animals—is fundamental to protecting us all from exploitation and the inconceivable.

These principles are timeless, perhaps because they reflect our biology. Our chemistry. The physics of who we are. Revered in ancient and modern philosophy and religious texts, from the East to the West, they are transmitted across generations and distinguishable from community-specific norms of cultural and religious institutions. They form the foundation for liberal democracies across the globe. In a cycle of struggle and renewal, they represent the idea of America—what we haven't yet accomplished but which drives us. Gradually we are beginning to understand how these principles also form the basis for nonhuman societies.

The pillars found in Phoenix Zones offer even more. They can move us beyond fallible intuitions and incomplete solutions that appear as optical delusions. As the building blocks for an ethical code, these principles can help us correct underlying prejudices and overcome gaps in empathy. We can build a new moral identity, one that is based on our commonalities rather than our differences. And they could even help us solve some of the biggest ethical and political challenges of our time—from methodical cruelty and abuse to mounting violence and terror.

But the best solutions aren't always the easiest. They are often complex, devoid of delusion and inconsistency.

THE OPTICAL DELUSION

Beginning in 2006, a small group of health professionals convened to stimulate interdisciplinary collaboration in all aspects of health care for humans and animals. Within a few years, they had launched a movement called the One Health Initiative.[3] As a worldwide effort, it aims to forge cooperation among human and animal health professionals, as well as government agencies like the Centers for Disease Control and Prevention and the US Department of Agriculture. The initiative is based on a concept dating back to ancient times, when physicians like Hippocrates stressed the connection between health and the environment. Its mission is to raise awareness about the inextricable link between human and animal well-being and to inspire action accordingly.[4]

The connections are profound. As one example, in the past three decades, world meat production has increased more than ten times the population growth rate—resulting in billions of animals being deprived of their dignity and sovereignty. Over time, the shift from plant-based to meat-based diets has contributed to rising human disease rates in both wealthy and developing nations.[5] Consequently, adults and children suffer from debilitating diseases like obesity, cancer, heart disease, and diabetes. Meat production also contributes to the spread of emerging and reemerging pathogens, which threaten the health of people and animals.[6] More than half of the infectious diseases found in humans are spread from animals. Crowded, unhygienic conditions in industrial farms, one of the most dangerous work environments for humans, increase the risk of transmitting these diseases.

Coincidentally, the rise in meat production has led to climate change, massive water pollution, soil erosion, and the destruction of rain forests and the animals who once lived there. According to an analysis by the late Robert Goodland, a senior environmental advisor to the World Bank, and the International Finance Corporation's Jeff Anhang, more than half of the greenhouse gases responsible for climate change stem from the meat and dairy industry.[7] Climate change is tied to numerous health costs to humans and animals—from drought and famine to vector-borne diseases like malaria.[8] People living in poverty, with little access to food or water, are at the greatest risk.

I have treated these communicable and noncommunicable conditions in the United States and abroad. As a public health professional, I have advised health policy leaders on how to address the incredible challenge of combating these illnesses effectively even with limited resources. I've met the animals whose lives were compromised by laws and economic incentives, the structural forms of violence that propelled their exploitation.

Unfortunately, many of the projects inspired by the One Health Initiative do not address the deepest links between human and animal well-being. Instead, they rely on modifications to meat and dairy production practices or comparative disease research involving animal experimentation. In these cases, animal suffering is discounted for human interests, an example of the ethical gerrymandering that legal scholar Ani Satz cautions against. Guiding principles like liberty and justice are applied haphazardly, or not at all. Such shortsighted approaches reflect deficiencies in empathy for animals like Farm Sanctuary's Julia, and they ignore potential social, health, and environmental consequences for people. In 2007, the South African philosopher David Benatar wrote about this problem.[9] At the time, avian influenza, one of many animal-borne diseases, threatened to become a pandemic. Public health efforts focused on vaccine development and killing birds. Benatar wondered why public health officials were looking for proximate rather than root causes of the problem:

> It is curious, therefore, that changing the way humans treat animals—most basically, ceasing to eat them or, at the very least, radically limiting the quantity of them that are eaten—is largely off the radar as a significant preventive measure. . . . Indeed, the curative and many of the preventive measures on which humans focus are ones that often involve further suffering and death for animals. For example, because humans have contracted diseases from maltreating animals, others then experiment on animals in a bid to find either a vaccine or a cure for the diseases that result from the maltreatment. Although these medical interventions are being developed, millions of animals are culled, often painfully, in the hope of preventing imminent disease or epidemic in humans. . . . When the (infected) chickens come home to roost, it may be another person, possibly from the next generation, who suffers or dies . . .[10]

This cycle, he notes, is not one of retribution but one of consequence. It is but one example of a failure to genuinely examine the connections between human and animal well-being. Such myopic solutions rely on the creation of simple independent categories—in this case, human and animal—and what Albert Einstein called an optical delusion. It is a delusion that derails us toward more suffering, more moral residual, and more violence that trickles down.

Consider for a moment if we instead addressed the truest links between health and violence across species. What if we infused the principles found in Phoenix Zones in all of our individual and collective choices—from business decisions to choices at the supermarket? How much further could we be in our progress? If we were to embrace all living creatures by expanding our circles of compassion, could we escape the optical delusion imprisoning us?

Yes, but only if we are willing to reimagine reality.

THE MIND'S ERRORS

In the 1970s, the Israeli psychologist Amos Tversky noted that "Reality is a cloud of possibility; not a point."[11] Tversky worked closely with another Israeli psychologist, Daniel Kahneman, as Michael Lewis describes in *The Undoing Project.*[12] Together, over the course of their careers, they challenged assumptions about human rationality and judgment. They showed how the human mind is fallible; in effect, how it creates optical delusions. Their conclusions have led to improvements in decision-making—from baseball, as described in Lewis's *Moneyball,* to data-driven evidence-based medicine. The title of Lewis's book is based upon a project that Tversky and Kahneman initiated to understand the rules of "undoing."[13] In an attempt to comprehend how the human mind undoes problems, they showed how we often select the most proximate answer to a problem, not necessarily the best solution.

A preference for simple solutions is the problem David Benatar underscored when he questioned the public health response to the threat of a bird flu epidemic. Public health officials attempted to address manifestations rather than the source of the problem. Rather than abandoning a system in which animals are crowded into cages to mass-produce meat and eggs, they tried to curb the effects of disease by killing birds and using other animals to develop a vaccine. In the end, the public health threat remained, even after an incredible cost to people and animals. The problem was only partially undone, if at all. The optical delusion that humans and animals are separate interfered with the selection of a comprehensive, consistent, and principled solution, one that would address the origin of the dilemma. An answer, even if insufficient, was offered, and few questioned its legitimacy. Gaps in empathy likely contributed to the shallow response and its acceptance.

THE EMPATHY GAP

During the course of their work, Tversky and Kahneman showed how an intolerance for uncertainty can cause reasoning errors founded on cognitive biases. Intolerance for uncertainty is linked to emotional regulation and safety and threat interpretation. To our brains, uncertainty can be read as a threat, perhaps due to the actions of the chemical dopamine on subcortical areas of the brain. It can lead to paralyzing fear or irrational actions, making our fallible minds susceptible to influence. Frequently enough, to avoid uncertainty, we choose either to do nothing, and become bystanders, or to automatically follow those who offer certainty. As a result, we may avoid novel situations perceived as risky, leading to confirmation of biases rather than correction of biases in the face of actual experience. An intolerance for uncertainty can therefore lead to a failure of empathy and what has been called the "bystander effect," a term popularized after the murder of Kitty Genovese in New York City.[14]

Gaps in empathy stem from an inability to recognize and respond effectively to the feelings of others.[15] They are at the heart of many polarizations in society, from racism to xenophobia to our treatment of animals. Most of us are capable of empathy, to varying degrees, but we distribute it inequitably. Our disproportionate appropriation of empathy has significant consequences—from our response to war to health disparities to the way we treat the human-animal bond.

In a 2016 *PBS News Hour* interview, Anne Barnard, a journalist for the *New York Times,* spoke about being stationed in Syria during its civil war.[16] She noted that—despite the recurring sad, shocking images of children harmed by the five-year-old conflict—fear and prejudice, predicated on uncertainty, became obstacles to empathy and subsequent action. On the whole, global citizens did not react altruistically to what the United Nations called the worst migration crisis since the Second World War. As tens of thousands of migrants and refugees from the Middle East and Africa fled war and persecution, nations tightened their borders. As fear of immigrants spread, more borders closed. Pictures of Alan Kurdi, whose two-year-old body washed up on a beach in Turkey, and five-year-old Omran Daqneesh, sitting stoically in an ambulance with a bloodied, ash-stained face, had only a momentary impact.[17] Around the same time, hate crimes against marginalized groups in the United States and Europe surged.[18] Individuals were targeted based on assumptions about their race or religion, among other characteristics. Some of the highest rates of harassment were reported in schools and universities, the supposed home of reason.[19] Still, other individuals and groups responded with compassion and courage. Some offered refugees temporary shelter and food. Others stood for nonviolent principles and against hate crimes.

Around the same time, other journalists pointed out how the public response to terrorist attacks differed depending on their geographic location.[20] Americans lent an outpouring of generous support after attacks in Paris and Brussels but by and large had a muted response for concurrent terror victims in Nigeria and Iraq. Other reporters commented on the racial empathy gap in the United States as the criminal justice and health care systems treated the lives of African Americans with less compassion.[21] Journalists struggled to explain the repeated killing of unarmed African American boys and men by law enforcement. Health disparities for African American, Latino, and economically disadvantaged citizens grew, rather than receded, by most measures.[22] Some reports showed that white people, including medical personnel, assumed people of color feel less pain.[23]

During the same period of time, animals suffered in cages until their deaths. There was an outcry of anger for Harambe, a gorilla who endured a life in confinement in the Cincinnati Zoo until he was shot and killed when a child fell into his enclosure. Meanwhile, some journalists pointed out that the same anger was lacking in other cases of animal suffering, particularly those killed for their meat, eggs, or milk.[24] Others revealed that the empathy for Harambe was not shown toward the family of the child who fell into his cage, and that the family experienced a backlash due to racism and sexism.[25] Biases were confirmed and empathy gaps widened. The violence and suffering continued.

We will never move past the optical delusion of separation and find true, holistic solutions if we do not close breaches in empathy.

CLOSING THE EMPATHY GAP WITH A CONSISTENT MORAL FRAMEWORK

We don't start out life with an empathy gap. Research has shown that it can emerge as early as elementary school and widen or narrow as we age. These changes are likely influenced by how we develop a group identity, which is one way our minds deal with uncertainty. The influence of group identification plays out across racial lines, ethnic delineations, and religious affiliations. It also affects how we treat nonhuman animals. Cognitive neuroscientist Emile Bruneau has even postulated that heightening empathy for one's own group could increase hostility toward others and lead to acts of violent extremism.[26] He has also hinted at a way to overcome disdain and apathy.

In a 2010 study, Bruneau arrived at an unexpected finding. To better understand the biases of Israeli and Arab adults in one of his Boston-based studies, he asked them to read short letters about the Middle East and rate how sensible the perspectives were. Israeli participants were more likely to have anti-Arab biases and rate the opinions of Arabs as unreasonable, and Arab participants were more likely to hold anti-Israeli biases and rate the views of Israelis as unreasonable. However, a small number of participants who held biases against the contrasting group were still able to identify with their perspectives. Each of the outlying study participants identified as a peace activist.[27] They each drew from a moral framework that seemingly transcended their biases. The optical delusion of separate categories began to clear.

Though the study was small, it raises a serious set of questions: Could a consistent principled framework ground us in a time of tremendous uncertainty? Help us overcome our biases, much like an evidence-based approach addresses cognitive biases about medical decisions? Could it help eliminate the empathy gap?

It could. Developing a moral identity built on adherence to an ethical code can broaden one's distribution of empathy. Imagining ourselves as caring individuals—shaped by core values like respect, kindness, and fairness—activates compassionate behavior. It can save us from indifference and inaction, as educational psychologist Michele Borba describes in *UnSelfie: Why Empathetic Kids Succeed in Our All-about-Me World*. As an example, she describes a boy who established a moral code when he first learned of Kitty Genovese's story.[28] Years later, as a pilot, he landed an incapacitated airplane on the Hudson River. Known as Captain Chesley "Sully" Sullenberger, he was the last to leave the plane. He walked repeatedly up and down the jet aisle before exiting, recalling his boyhood promise to leave no one behind. Borba found a similar pattern while visiting genocide memorials, which all noted people who refused to be bystanders. Social science studies support these anecdotes.[29] A strong moral identity can help guide future decisions—even after a plane crash or during and after genocide, as it did for my patient Benjamin in Rwanda.

But we need to apply the pillars found in Phoenix Zones consistently in order to effectively address the empathy gap and break free of the optical delusion imprisoning us.

As psychiatrist Sujatha Ramakrishna has advised, the ways children witness animals being treated also have a strong influence on whether they develop principled behaviors such as respect, sharing, and cooperation.[30] Even at an early age, children realize that consistency—the absence of contradictions—requires application of the same moral standards across the board unless two situations differ in meaningful ways. They seem to understand what ethicist Tom Beauchamp has written about for students of philosophy: we already recognize that animals have moral standing in our society. They have emotional and cognitive properties that qualify them for moral obligations and therefore rights and the consistent application of moral principles.[31] And they might also teach us how to achieve a more valuable group identity, one based on our shared capacity for vulnerability and resilience. Not what a woman means to a man, or a nonhuman animal to a human. But instead, how she can rise in and for the world around her.

A consistent and inclusive moral framework could even help us solve some of the most monumental challenges of our time. But we need to move quickly. We could be running out of time.

DREAMING BIGGER AND SCALING UP

In *Thank You for Being Late: An Optimist's Guide to Thriving in the Age of Accelerations,* Thomas Friedman chronicles the leviathan shifts reshaping our world. He argues that three forces are driving these changes. Beginning with a story about the invention of the first smartphone, Friedman illustrates the exponential rise in technologies since 2007, largely driven by Moore's law.[32] Human adaptability, Friedman warns, is not keeping pace with technological advances. At the same time, "the Market" is hastening, often through digitalization that drives "the globalization of friendships and finance, hate and exclusion" and more, "so that good and bad ideas can go viral and extinguish and manufacture prejudices much more quickly . . ."[33] These shifts introduce risk and opportunity. One clear danger, Friedman warns, is the effect that accelerations in technology and globalization have on climate change and biodiversity loss ("Mother Nature," the third force).[34]

In an age of accelerations, incremental change and step-by-step social progress will not be sufficient. We need to dream bigger and plan on a scale commensurate with what is at stake. Attempts to quantitatively reduce suffering—such as through engineering standards like cleaner, larger cages for animals—are respectable but inadequate. Everything we do to minimize suffering matters, but we cannot quantify someone else's qualitative experiences, particularly if we don't completely understand their life. We know from Phoenix Zones that recovery is made possible when barriers to freedom and the opportunity to thrive are removed. Ignoring a commitment to key principles leaves the door open to reverse engineering, to the detriment of people and animals. Industry can always come up with another way to enhance human interests while making animals suffer. Policymakers can manufacture ways to impede freedom and justice for other humans, such as through the creation of post-Reconstruction Jim Crow laws or voter suppression activities. Likewise, psychologists and officials can forever contrive enhanced interrogation techniques to circumvent international laws against torture.

Gradual changes are clearly important, but there is no proof that they are the best or only way to achieve social justice. On the contrary, some of the largest leaps in social progress have followed a steadfast dedication to key principles—and, of course, a lot of hard work. Certainly there's been the ultimate positive impact of some wars, technology, and market forces. But only through principles. Though the Second World War and the Civil War brought liberation, they were driven by a commitment to justice. Imagine if a universal commitment to love, tolerance, and sovereignty were in place before the Nazis gained power in Germany, or if America's founding fathers truly had a relentless pledge to freedom—for all, not just some. Would the Second World War or the Civil War have occurred? We almost never ask such questions in our undoing of historical events.

If we are interested in meaningful change, we should take Abraham Maslow's approach, which asks: What do we need to truly flourish as individuals and as a society? And further, who are we, and who do we want to become?

In December 2014, the Senate Intelligence Committee released a report on the CIA's use of torture tactics, which were largely driven by anxiety and insecurity, under the Bush administration. The tactics are believed to have fueled rather than deterred terrorism. Senator John McCain, a former prisoner of war and a torture survivor, has been one of the most vocal opponents of torture. Once the Senate report was released, he gave a speech on the Senate floor. He said:

> It is a thorough and thoughtful study of practices that I believe not only failed their purpose—to secure actionable intelligence to prevent further attacks on the U.S. and our allies—but actually damaged our security interests, as well as our reputation as a force for good in the world. . . . I know from personal experience that the abuse of prisoners will produce more bad than good intelligence. . . . But, in the end, torture's failure to serve its intended purpose isn't the main reason to oppose its use. I have often said, and will always maintain, that this question isn't about our enemies; it's about us. It's about who we were, who we are, and who we aspire to be.[35]

The definition of "terror" is extreme fear. "Terrorism" is defined as violence committed by a person, group, or government in order to frighten people and achieve a political goal. If we succumb to fear and forego our principles, terror wins. Now, we face the challenge of creating more sanctuary in the world without being seduced by fear. Will it immobilize us or will it become part of a broader context that informs more effective action?

By infusing values such as love and respect for personal sovereignty, dignity, and justice more deeply in society, we could dismantle cultural forces that foster and normalize violence—from industrialized cruelty to terrorism. Counterterrorism experts have already begun to focus on the problem of group identity and the importance of moral identity in prevention and deradicalization efforts.[36] In an effort to prevent and address conflict, neuroscientists and psychologists have stumbled across another finding: narratives can shift norms. The more ubiquitous values are, including in our treatment of animals, the more applicable they are. And the more substantive they are, with clear consequences and functions, the more likely they are to be adopted as normative behaviors. A set of principles, absent contradictions, can help us overcome erroneous intuitive judgments and the resulting violence. Moral progress isn't easy, but in many ways is straightforward, if we have the moral clarity and conviction required to make substantial advancements.

Our fate, and our quest for sanctuary, is largely dependent on whether we commit to a common set of nonviolent principles. The capacity to achieve peace depends on whether we invest time, money, and action accordingly. We need not sacrifice one principle for another. On the contrary, we need an uncompromising commitment to all of the values found in Phoenix Zones.

After all, what does freedom mean if not for sovereignty? Without justice, how would we guard freedom and sovereignty while guiding love and tolerance? Absent love, how would we temper justice? Only with hope and opportunity can we aspire for these ideals. And without respect for the inherent worth of each individual, where do we even begin?

But we too seldom consider how these principles apply to our daily lives and choices. Under the threat of uncertainty, we could move further away from them. History has shown how a failure to consistently apply principles like liberty and justice can result in a loss of their merit. Now more than ever, with consistency and courage, we need to build Phoenix Zones that reinforce these values.

10 : OPENINGS TO THE IMPOSSIBLE

The men and women who have every reason to despair, but don't,
may have the most to teach us, not only about how to hold true to our beliefs, but
about how such a life can bring about seemingly impossible social change.
— PAUL ROGAT LOEB, *The Impossible Will Take a Little While*

In 2015, a battle of ideas broke out in *The Guardian*, widening a divide between two schools of thought.[1] Author John Gray criticized a belief shared by psychologist Steven Pinker and philosopher Peter Singer, among others, that violence and war are decreasing. Gray defended his own view that we as humans have made little to no significant moral progress over multiple millennia.

In *The Better Angels of Our Nature,* Pinker contends that human beings are becoming less violent and more altruistic, an argument he bases on specific statistics and historical events.[2] Drawing on the German sociologist Norbert Elias's twentieth-century theory of civilizing processes, Pinker suggests that we are gradually conquering violence.[3] To Pinker, the "better angels" subduing our "inner demons" arise from reason, as well as the spread of commerce and technology and changes in values influenced by the Enlightenment.[4] Consequently, he argues, norms affecting people and animals have changed for the better. Over the latter half of the twentieth century, more attention was focused on civil rights and women's rights. Children's rights gained traction. Since the 1970s, gay rights and animal rights have received more consideration. Pinker refers to these advances as "Rights Revolutions."[5]

Gray claims that Pinker's conclusions rely on a selective memory of history. Human rights ideals aren't novel, just as ideas about animal rights aren't new. Only a couple of centuries after Cyrus the Great released the slaves in ancient Persia and established religious and racial equality, Aristotle's pupil Theophrastus wrote about the importance of treating animals justly.[6] He wasn't alone. Some of the earliest objections to the use of animals for food and experiments are found in ancient writings. In the early part of the first millennium, the Neoplatonist philosopher Porphyry argued against treating animals as if they were mere objects for our use.[7] Long since then, animals have become captive to industrial cruelty and subject to what author Elizabeth Kolbert terms the "sixth extinction."[8] Commerce and technology have not played a benign role in these cases.

In Gray's view, Pinker conveniently overlooks methodical violence promoted during the Enlightenment period and since, while demeaning indigenous cultures and downplaying the violence perpetuated by wealthy nations:

> There is something repellently absurd in the notion that war is a vice of "backward" peoples. Destroying some of the most refined civilizations that have ever existed, the wars that ravaged southeast Asia in the second world war and the decades that followed were the work of colonial powers. One of the causes of the genocide in Rwanda was the segregation of the population by German and Belgian imperialism. Unending war in the Congo has been fuelled by western demand for the country's natural resources. If violence has dwindled in advanced societies, one reason may be that they have exported it.[9]

Despite Pinker's positive account, human atrocities continue. In some places, they've intensified. It's impossible to ignore the influence of authoritarian regimes and mass violence that plagues much of the globe, or to assume that peaceful trends will necessarily continue in other parts. And despite Gray's claims and the flood of suffering around the world, there *are* clear signs of moral progress. International rules now prohibit violence committed disproportionately against women and girls during war, particularly rape and sexual slavery. Only recently has rape become a crime within international law; the earliest prohibitions of rape existed to protect women by virtue of their status as the property of men. Measured strides toward racial equity in the United States have been made. Addressing the legacy of slavery and Jim Crow laws, including inequitable mass incarceration practices, has garnered more support.[10] In Europe and North America, laws increasingly forbid discrimination based on sexual orientation. Marriage equality is one area of achievement. Animals have also seen some gains—circuses are closing down, animals' lives are receiving more attention, and laws to protect them are evolving.

For everyone who has benefited, these advances matter. Yet for anyone still affected by subjugation or violence, Pinker's and Gray's perspectives are seemingly inconsequential. Lost in their argument is what we can do as individuals and as a society to effect change through personal choice, grassroots action, and policy change. We can become neither complacent because of Pinker nor discouraged because of Gray. Going forward, we will be responsible for the moral residual we create and for the empathy we distribute.

But in an age of accelerations, when the accumulation of historical and ongoing oppression weighs heavily upon us, how can we rise up to face the incredible task of meaningful, seemingly impossible, widespread change? Those who have every reason to despair, but don't, provide instruction. From them, we can learn how to leave our silos, find common ground, and discover unimaginable rewards among incredible costs.

FROM AN INCREDIBLE COST TO AN UNIMAGINABLE REWARD

"It has been both the most horrible and most gratifying experience of my life."[11]

That is how one clinician responded when asked to reflect on her experience providing forensic examinations to torture survivors. The question was one of several in a paper I coauthored with my colleague Ranit Mishori, a family physician and former war correspondent.[12] It was the first published study to evaluate the motivations and experiences of health professionals who provide asylum evaluations. The doctor's statement echoed another sentiment—from a different time and place—when a veterinarian who cared for abused animals told me, "They've taught me so much. Some things I'd rather not know." Both health professionals were grappling with the costs of compassion.

In 2006, animal law professor Taime Bryant warned that vicarious trauma has the potential to undermine social progress.[13] Tragedies, even those we don't experience directly, stay with us. Our bodies absorb the visceral costs of compassion in an imperfect world. Others' memories become our memories. Their pain, our own. Intellectual inflexibility, cynicism, and distrust of others—compassion fatigue—can compromise efforts to raise awareness and summon action.

Frequent contact with violent images and experiences exposes advocates to the risk for their own posttraumatic stress symptoms. Journalists face similar threats to their mental health, and in an era of user-generated content and widespread media dissemination, every empathic person is at risk. Virtually, we are in dozens of places at once, attuned to the vulnerability and suffering of people and animals living all over the world. Through social media, we are on the front lines of war with a rape survivor. On the nightly news, we are looking in on the life of an abused child. Or some of us, perusing a magazine, find ourselves in a factory farm or at the site of a gorilla's murder by poachers. And even with all we already see, hear, and know, learning about more suffering often opens old wounds, leaving us with fresh and heavier pain. Impunity and dismissive attitudes about suffering—public gaps in empathy—increase the risk for this sort of vicarious trauma.

But stories of struggle and triumph can also empower us, particularly if we are part of them. As another clinician in our study said, "I am inspired, saddened, enraged by what I hear and see—all of which deepens my commitment to the work."[14] Yet another said, "Their stories have changed my life."[15] Her statement became the title of our paper, since it reflected the sentiments of so many other medical professionals in our survey.

In our study, many described experiencing personal growth after helping torture survivors. It deepened their appreciation for living in a free and tolerant society. When asked about their motivations, some reflected on an expanded sense of purpose—one of the psychological needs Abraham Maslow articulated decades ago. Just one year after our study was published, another team of scholars printed their own findings on human rights professionals working around the world.[16] They aimed to understand the impact of human rights work on the mental health of advocates. Though reports of PTSD and depression symptoms were common, rates of resilience (absence or resolution of symptoms) among professionals were also high. The work itself, and a sense of achievement, inspired hope and motivation. As Jerome Groopman suggests in *The Anatomy of Hope*, a cycle of positive beliefs and expectations can shield us from a sense of helplessness, to which all of us are susceptible. It can give us the strength to change the world we live in.

LEAVING THE SILOS

In her article on compassion fatigue in activists, Bryant notes the difficulty of working within a single social movement, let alone more than one:

> When activists, *in any social justice movement,* must be wholly dedicated to speaking truths that others will not or cannot hear, it is a gargantuan task for them to take on all of the forms of oppression intertwined with the truths they speak. . . . Each social justice movement participant is fighting just to be heard about the particular manifestation of violence he or she sees most clearly. . . . Addressing both [human and animal cruelty] involves even more difficult work: reconceptualizing or dismantling the entire enterprise that oppresses both animals and people.[17]

Given the enormous task of taking on multiple sources of oppression, it's unsurprising that social justice movements today work in separate silos. Rather than working toward the full protection of people and animals, many have subscribed to culturally constructed identities—group identities—grounded in restricted independent categories. In some important instances, this approach has resulted in victories for those working in the fields of international human rights, civil rights, women's rights, and animal rights, among others. It has created space for marginalized individuals where there once was none. Nevertheless, despite strides in some areas, the road to social justice is still quite long. Activists working on behalf of animals wage many of the same battles against cruelty that advocates like Henry Bergh of the ASPCA did in the late nineteenth century. Notwithstanding the advances that Bergh and others made for children's rights, advocates must still fight physical, sexual, and emotional abuse perpetuated against children. And even after years of campaigns against the use of torture as a method of social control, some are still arguing for its usefulness.

We have not yet successfully infused the principles of Phoenix Zones—those that are necessary for recovery—into society. But we can.

Now is the time for more inclusive progress: a unified effort to address the injustices that have led to all forms of marginalization and required the erection of silos in the first place.

When I reflect upon those who have created Phoenix Zones, I am struck by their uniform dedication to people and animals. J. B. and Diana are devoted to the Cle Elum chimpanzees, but they have also chosen to work with and on behalf of vulnerable children. In Brazil, Kat and Scott treat their employees, volunteers, and members of the community with the same respect they give the elephants. At the Warriors and Wolves project, Lorin and Matt maintain a loving commitment to veterans and animals, and Lorin remains loyal to troubled children and adults in her clinical psychology practice. In Oklahoma, Penny felt compelled to find shelter and justice for young people and their companion animals. Neema has demanded opportunities for women, girls, and the animals living around her in Congo. In California and upstate New York, Susie works hard to create sanctuary for abused farm animals, but she is as present for the young women and men who work at the sanctuary. The guardians of Phoenix Zones aren't victim to the optical delusion of separation; to them, the principles of sanctuary are ubiquitous. They are generous with their empathy.

Those working at the intersections of social justice, like writer and scholar A. Breeze Harper, also provide guidance for how to leave our silos. Based on her own and others' experiences related to categorizations such as race, gender, orientation, and species, Harper explores systems of oppression and power.[18] As she and others show, social movements constructed on the basis of categorical representation need not work in opposition, nor in competition.[19] Our futures are connected. A holistic anti-oppression framework reflective of key principles reinforces shared goals. But first we need to be willing to leave our silos, and to disrupt the ground they are built upon.

DISRUPTING THE GROUND AND ABOLISHING THE VECTOR

In public health, there are distinctions between the control, elimination, and eradication of a disease. Control of a disease implies a reduction in the rate of spread. Elimination indicates that transmission has stopped in a specific region but not worldwide. Reaching zero—eradication—is always the end goal, though it is often more difficult, at least in the short term, than controlling a disease. To be fully eradicated, a disease must be eliminated everywhere throughout the world. For many, eradication of disease, especially the disease of violence, seems impossible. But we should always work toward eradication. If we don't, the risk for transmission persists. The contagious particle remains.

Within preventive medicine, a medical field focused on public health interventions, there are three types of prevention. Primary prevention focuses on addressing the root causes of disease and preventing it before it occurs; for example, avoiding exposure or abolishing the cause of disease. Secondary prevention focuses on catching a disease before it spreads, such as through tests that screen for illness. Tertiary prevention aims to limit the consequences of a disease; for example, through rehabilitation programs. Though we employ all of these prevention strategies in medicine, the best approach is primary prevention: averting illness before it occurs.

To eradicate the contagion of violence, we need a primary prevention strategy to reconceptualize the entire enterprise Bryant writes about: to abolish the oppression that serves as violence's vector. To undo the cultural, political, and economic structures that allow violence to seep down into society. Already, there are opportunities to disrupt existing institutions and compassionately reconfigure the accelerating enterprise we live in. One opportunity is through the One Health Initiative; specifically, what David Benatar suggested as a fuller, more thoughtful and compassionate public health response to prevent diseases like bird flu. Another example is found in the field of education, where many solutions begin.

In 1996, educators Zoe Weil and Rae Sikora cofounded the Institute for Humane Education. Under the notion that "the world becomes what we teach," it focuses on interconnected issues that impact all life.[20] In recent years, the institute and its alumni have turned to developing an open-source, multidisciplinary "solutionary" curriculum for children and adolescents in public and private schools. Using what they have learned in their core classes, students are asked to solve some of the most pressing global and local problems, and to address the underlying causes of each dilemma without harming people, animals, or the environment. One large-scale, international effort asks students to select problems they consider important, and to then work together to create humane and just solutions. Students learn how to apply a central ethic consistently, even among real-world challenges. In regional and national forums, students present their solutions to stakeholders who can help further craft and implement their ideas. Efforts like these could provide an antidote to what Michele Borba calls the self-absorptive "Selfie Syndrome."[21]

Through humane education, kids learn critical-thinking and consensus-building skills, and they acquire perspective-taking skills, which can help narrow their gaps in empathy. Perspective taking—understanding another's thoughts and needs—helps eliminate unconscious biases, especially if it reveals commonality. Individuals who identify with others are more likely to refuse to be bystanders, and tolerance for uncertainty can lead to less fear and more opportunities to correct cognitive biases. In her book, Borba shows how kids can develop the ability to understand and act on the feelings of others, even how an infant can teach third graders to identify signs of distress.[22] Techniques like these have been incorporated into educational curricula in Denmark,[23] and these methods resemble how animals can also teach children empathy.

As studies have shown, children acquire a sense of justice at a young age. Values like empathy can serve as a lens through which children can eventually make critical assessments of the accelerating technologies and markets they will inherit.

There are already signs that markets are being influenced by compassion and other values found in Phoenix Zones. In 2016, Wayne Pacelle, president and CEO of the Humane Society of the United States, published *The Humane Economy*.[24] In it, he shows how business leaders, along with scientists and philanthropists, are disrupting industries built on exploitation. Consumers who care about freedom (and not just their own) drive markets by choosing what to buy, or what not to buy. From agriculture to fashion, Pacelle describes how executives and shoppers increasingly take into account animals' interests, in addition to human interests. By figuring out how to invest in a more just future, they are what humane educators might call "solutionaries."

The media outlets we support also influence many of our decisions as consumers and citizens. Filled with words and images, they shape our biases and how we distribute empathy. Violent or disrespectful images can condition children's brains to tolerate and even perpetuate cruelty, whereas caring stories and visuals can help spread kindness. Media can include or exclude, illuminate or obliterate, those who are not in positions of power.[25] Very often, the media is the only connection people have with many species of animals. Still, today this dichotomy also resembles how people of different backgrounds interact with one another. As a result, media depictions of marginalized people and animals are critical to creating or maintaining an empathetic sense of them.[26]

Elsewhere there are more examples of prevention strategies to tackle sources of violence. Working on the local, national, and international levels, the National Link Coalition focuses on education and training, as well as network building, to highlight "the link" between cruelty to people and animals.[27] It also works to strengthen laws and reporting requirements. In some places, the law has already begun to recognize the link between violence directed at people and animals. In 2016, the Federal Bureau of Investigation began collecting data on animal cruelty through its National-Incident-Based Reporting System.[28] Like never before, these abuses are now tallied right along with arson, assault, and homicide. Still, many animals are ignored in this initiative, reflecting the need for efforts like those of Nonhuman Rights Project lawyer Steven Wise to establish animal rights jurisprudence—a change in the way the law sees animals. Such a change could benefit both people and animals by addressing all, not just some, roots of violence. Ideally, law enforcement efforts like these won't merely record abuse but will prevent other abuses from occurring.

Internationally, there have also been significant strides to address the tie between violence involving people and animals. For example, during the Obama administration, Congress and the Departments of State and Justice prioritized related anti-terrorism and anti-poaching activities.[29] It's clear why. Poaching rhinos and elephants, among other animals, has been credibly linked to pockets of terrorism around the globe, from the Islamic State of Iraq and Syria to Boko Haram in Nigeria. Since these are borderless crimes linking people and animals, the time has likely come for former United Nations ambassador Muhamed Sacirbey's suggestion to appoint a global ambassador for animals.

Within nations, it could also be time for cabinet appointments focused on the primary prevention of violence, including "the link," and reconfiguring our institutions accordingly. Already, more and more political parties across the globe acknowledge the importance of considering humans *and* animals in rulemaking. If politics sets a precedent for future legislative change, change could happen sooner rather than later.

What an exciting challenge and opportunity—to create solutions devoid of optical delusion, ethical gerrymandering, and the creation of moral residual—to rebuild institutions with the pillars of Phoenix Zones.

WE CAN CHANGE THE SYSTEM

Working around the globe, I see how my friends and colleagues carry the weight of social and cultural reform. I've met far too many men and women who have been tortured or humiliated for opposing attempts by their government to limit the rights of others. Despite the uncertainty around them, they, like the creators of Phoenix Zones, find courage in vulnerability. They know that all institutions depend upon the people who guide and comprise them. As my friend Neema says, "If we change, the system will change around us."

There are opportunities everywhere we turn. We cannot assume someone else will come along and respond. Someone has: you and I, who can be "Upstanders" rather than bystanders.[30] One person, if willing, can change the world of one individual. For one, we can each create sanctuary. The dignity shown toward a person who has been discounted matters to that one person. A refusal to stand by an assault—on anyone—is consequential to that one being. Offering a loving forever home to an animal at a shelter or on the street is significant for that one animal. The opportunity given to a child is important to that one child.

But our influence doesn't stop there. We have the power to affect other people and animals, including those we will never meet, in countless ways. For example, decisions about what, or who, we do and don't put on our plates matter. The lived histories of all of our purchases—the people and animals required to create them—are real. Our choices online, or at the market, are meaningful to them. We can drive a humane economy. And we can disable exploitative institutions by holding firm to the principles of Phoenix Zones—by refusing to support industries that violate the freedom and sovereignty of humans and animals or deny them justice and the opportunity to live up to their full potential. Instead, we can support businesses and activities that foster love and dignity.

Even our language matters. Changing "they" to "we" and choosing words other than "it" to describe other sentient beings can improve the way we treat individuals other than ourselves. Removing words that are meant to denigrate others from our vocabulary, including those that refer to animals, is also important. As my friend, media scholar and journalism professor Debra Merskin, once told me, "If, for example, the worst thing one can be called is an 'animal' or the most horrible experience is to be treated like one, where does that leave animals?" And, as *Eternal Treblinka*'s Charles Patterson also cautions, where does that leave the humans who are compared to them?

All of our decisions have consequences. From our language, to our choices as citizens and consumers, to our judgments about how we guide and educate children, we are shaping the world. Preventing violence or spreading it. Dismantling an enterprise of oppression or bolstering it.

A SENSITIVE DEPENDENCE ON INITIAL CONDITIONS

As mathematicians and scientists have shown, there is order in the chaos of the universe. Tiny differences in input can result in overwhelming differences in output. One example is the butterfly effect. The term, coined by physicist Edward Lorenz, is derived from the metaphorical example of how a butterfly's flapping wings can influence the path and force of a hurricane weeks later.[31] Similarly, though individuals and small groups of people can do a great deal of damage, the converse is also true. Individuals and small groups of people can work wonders.

In chaos theory, the butterfly effect signals a "sensitive dependence on initial conditions."[32] If the initial conditions are principles, we need to start there, apply them with consistency, and build up to rise up. While further progress still needs to be made on behalf of humans, the health and well-being of people and animals are not mutually exclusive. Efforts to suggest otherwise reflect a lack of imagination; they are blinded by the delusion Einstein forewarned of. There is common ground occupied by those working on behalf of people and animals—both because of the shared potential for suffering and because many solutions to successfully combat violence are universal.

Many of the ways animals are used in society are reminiscent of how humans have been and continue to be mistreated. Excluding animals from key protections opens the door to the subjugation, discrimination, and abuse of humans. It undermines the fundamental principles on which protections for humans are based. For as long as our institutions are built on oppression and exploitation, we will not find the freedom or justice we seek. Our love and opportunities will be limited.

But we cannot mistake what others will call the impossible for the impassible. Every victory for the beleaguered was once considered impossible. Up against the fear of uncertainty, and devastating defeats, significant progress appears implausible at times. But, as many engaged in the ongoing struggle know, it is not only possible but necessary. Obstinate obstacles to social justice are self-limiting. Wrongs can crumble—when freedom breaks through prisons; sovereignty halts trespasses; love trumps hate; justice overpowers inequity; hope surfaces from despair; and dignity transcends degradation.

Each one of us can make a difference in the life of one and the lives of many. Though it often requires courage, the cost is worth the reward. And by working together, we can turn vulnerability into strength, and desperation into hope—the breath of fire that fuels the Phoenix.

ACKNOWLEDGMENTS

This book was made possible only through the help and support of a number of extraordinary individuals. I have to begin with those who contributed most to the book, particularly the people and animals who inspired and filled its pages. To those who remain anonymous but gave immensely by sharing their brilliant lives and stories with me throughout the years, I am eternally thankful for the lessons they have passed along and for reviewing the material for accuracy. I am equally indebted to the people who are featured in this book for their efforts to create meaningful sanctuary for others: Kat and Scott Blais, Susie Coston, Diana Goodrich, J. B. Mulcahy, Neema Namadamu, Penny Reynolds, Amber Richardson, and Matthew Simmons. The time they took out of their busy schedules and important work to share stories and review material was invaluable. Lorin Lindner deserves special mention for her careful read of each chapter and sincere encouragement, despite juggling clinical and many other responsibilities. I'm extremely grateful to her and to the others for their openness, generosity, friendship, and most notably their remarkable efforts to improve the world.

My admiration and gratitude also extend to the staff at Bread for the City Medical Center, Chimpanzee Sanctuary Northwest, Farm Sanctuary, Global Elephant Sanctuary, HealthRight International, Hero Women Rising, Lockwood Animal Rescue Center, and Physicians for Human Rights. Their dedication is inspirational.

Additionally, I want to express my heartfelt thanks to others who talked with me or shared material for the book, many of whom are also valued friends: Kate Aubry, Laura Bonar, Alka Chandna, Debra Durham, Jenny Edwards, Bruce Friedrich, Debra Merskin, Zoe Weil, and Steven Wise. I am particularly appreciative of Wes and Margo Savoy, who also read the manuscript and provided important feedback.

Thank you to my editor, Christie Henry, whose enthusiasm and thoughtful guidance have been immeasurably helpful. Often she understood what I was trying to accomplish before I was able to articulate it, and she gently and unassumingly steered me in a better direction. I also owe sincere thanks to my agent, Barbara Braun, for her faith, representation, and direction, and to her associate John F. Baker. Additionally, I appreciate the editorial, marketing, and administrative staff at the University of Chicago Press for their professionalism, attention to detail, and the optimistic spirit in which they work. Special thanks to Marianne Tatom and Yvonne Zipter for their careful review and copyediting of the manuscript.

Several friends, colleagues, and family members read the manuscript, or earlier versions of the manuscript, and provided me with valuable suggestions. They include Mike Anastario, Melanie Kaplan, Allan Kornberg, Thomas McHale, and Rachel Wallach. Lauren Choplin deserves distinct appreciation for providing ongoing instrumental editorial guidance and support. Likewise, in addition to providing helpful thoughts on the final manuscript, Marc Bekoff has been a positive and energetic champion for this project since its inception. John Gluck has too offered wise advice and careful review. The many thoughtful conversations I've had with John over the years sparked or advanced many of the ideas included in these pages.

My appreciation also extends to others whose work, exchanges, and collaboration have challenged and advanced my understanding of subjects touched upon in this book: among many others, Randi Abramson, Neal Barnard, Tom Beauchamp, Martin Brüne, Chong Choe-Smith, David DeGrazia, Agustin Fuentes, Lori Gruen, Lise Holst, Jane Johnson, Jeffrey Kahn, Lori Marino, Angela Martin, Katalin Roth, and David Wendler. I'm equally grateful to others who have provided specific help and advice for this project along the way: Anthony Bellotti, Christina Deptula, Lawrence Deyton, Eric Engles, Mary Esselman, Jane Friedman, and Ranit Mishori. Many thanks to the Arcus Foundation and the National Science Foundation for supporting some of my work described in these pages.

I am obliged to two close friends: Ayanna Buckner for her always forthright and thoughtful advice, and Sonia Silva for her steadfast support and candor. Through both of them, I've acquired a deeper understanding of the difficulties and inequities in the world and what we can do to overcome them.

To my parents, for providing kind and honest feedback and endless encouragement, and more importantly an early tolerance for dissent, as well as an insistence that their children develop their own moral courage. I am as grateful to two of my lifelong friends, my sister Carin and my brother Kevin, for an infinite dialogue of constructive ideas over the years—from our gardens in Oklahoma to the sand dunes of Death Valley.

Finally, I want to thank my husband, Nik, for every comment on every work in progress, as well as his love, unwavering support, and patience—not to mention each laugh, careful argument, and adventure. But I am especially thankful for his sense of justice, quiet generosity, and the humble way he acknowledges the worth of others. And for sharing our home with Buster, Champ, Charlie, Lucy, Stella, Sugar, and Wagner. Though they will never read this, they have helped me develop many of the sentiments in this book. I hope they have felt an infusion of freedom, sovereignty, love, justice, hope, opportunity, and dignity in our lives together, even as I was learning how to do better.

RESOURCES

ANIMAL SANCTUARIES

There are many legitimate sanctuaries for animals. However, many organizations that claim to be sanctuaries do not apply the principles of sanctuary discussed in this book.

The **Global Federation of Animal Sanctuaries** is a resource for determining whether an organization constitutes a legitimate sanctuary: http://www.sanctuaryfederation.org/gfas/.

A list of the sanctuaries featured in this book follows. Generally, these sanctuaries are not open to the public since they exist to serve the needs of those who live there, but more information about them can be found through the links below.

Chimpanzee Sanctuary Northwest: https://www.chimpsanctuarynw.org.
Farm Sanctuary: https://www.farmsanctuary.org.
Global Sanctuary for Elephants: http://www.globalelephants.org.
Lockwood Animal Rescue Center/Warriors and Wolves Program: http://lockwoodarc.org/warriors-wolves/.
Serenity Park: http://parrotcare.org.

OTHER RESOURCES

The following organizations and websites provide further information on topics covered in this book.

Animals and Media is a media resource including style guidelines for practitioners in the professions of journalism, entertainment media, advertising, and public relations. It offers concrete guidance for how to cover and represent nonhuman animals in a fair and respectful manner in accordance with professional ethical principles: http://www.animalsandmedia.org.

BeFreegle Foundation is a nonprofit organization dedicated to providing sanctuary and rehabilitation to dogs who have been used in research: http://befreeglefoundation.org.

The Chandler Edwards Group has trained hundreds of social workers, veterinarians, law enforcement officers, and prosecutors on how to spot signs of sexual abuse in animals. While maintaining a National-Incident-Based Reporting System–level database of offenders, its staff has supported legislation in multiple states and foreign nations to ensure that sexual abuse involving animals is counted as a crime. It has assisted in investigations resulting in convictions, while encouraging ongoing research on the link between sexual violence crimes against animals and children: http://www.mjennyedwards.com.

Dylan's Wings of Change is a foundation established to honor the memory of Dylan Hockley, who was killed at Sandy Hook Elementary School on December 14, 2012. The organization was cofounded by his parents, and its mission is to help children with autism and related conditions achieve their full potential: http://www.dylanswingsofchange.org.

The Freedom Fund is a philanthropic initiative designed to bring strategic and financial resources to the fight against modern slavery: http://freedomfund.org.

Girls Not Brides is a global partnership of civil society organizations committed to ending child marriage and enabling girls to fulfill their potential: http://www.girlsnotbrides.org.

Global Network of Women Peacebuilders is a program of the International Civil Society Action Network (ICAN) and an international coalition of women's groups and other civil society organizations that are actively involved in advocacy and action for the implementation of the Security Council resolutions on women and peace and security: http://gnwp.org.

The Good Food Institute is an example of an organization driving a humane economy. It is a nonprofit organization that promotes plant-based and "clean" alternatives to conventional meat, dairy products, and eggs. Composed of a team of scientists and policy and business experts, it focuses on using markets and technology to compete with animal-based products to address intersections between human health, animal welfare, global poverty, and environmental degradation: http://www.gfi.org.

HealthRight International is a nonprofit organization that empowers marginalized communities to live healthy lives. Its Human Rights Clinic trains and deploys physicians and mental health professionals to provide forensic examinations and testimony for immigrants seeking relief in the United States: https://healthright.org/forensic-evaluation-services/.

Hero Women Rising is a nonprofit organization that works to support the grassroots empowerment of women in the Democratic Republic of Congo and beyond: http://www.herowomenrising.org.

Institute for Humane Education is a nonprofit institute that aims to create a just, humane, and sustainable world through education: https://humaneeducation.org.

The Last 1,000 is a web-based resource chronicling the journey of chimpanzees from laboratories to sanctuaries in the United States: http://last1000chimps.com.

National Link Coalition serves as a national resource on the link between animal abuse and human violence. It is an informal, multidisciplinary collaborative network of individuals and organizations in human services and animal welfare and focuses on research, public policy, programming, and community awareness: http://nationallinkcoalition.org.

Nonhuman Rights Project is a nonprofit organization that aims to attain legal personhood for nonhuman animals by operating within the reigning legal paradigms: http://www.nonhumanrightsproject.org.

Office of the United Nations High Commissioner for Refugees is a global organization that protects and assists refugees, forcibly displaced communities, and stateless people: http://www.unhcr.org/en-us/.

Parrots Forever Sanctuary & Rescue Foundation is a nonprofit volunteer-based organization dedicated to providing rescue and sanctuary for unwanted, neglected, or abused companion parrots. It also provides education and support to the public and to people who care for parrots: http://parrotsforever.com.

Physicians for Human Rights is a nonprofit organization that uses medicine and science to document and call attention to mass atrocities and human rights violations, including torture, sexual violence, and the persecution of health workers: http://physiciansforhumanrights.org.

Sandy Hook Promise is a nonprofit organization that aims to build a national movement of parents, schools, and community organizations engaged and empowered to deliver gun violence prevention programs and to mobilize for the passage of relevant state and national policy. It was cofounded by and is co-led by several family members whose loved ones were killed at Sandy Hook Elementary School: http://www.sandyhookpromise.org.

Sheltering Animals & Families Together is a global initiative guiding domestic violence shelters on how to house families together with their companion animals: http://alliephillips.com/saf-tprogram/.

Virunga Alliance aims to foster peace and prosperity through responsible economic development for the four million people who live within a day's walk of the park's borders: https://virunga.org/virunga-alliance/.

NOTES

INTRODUCTION

1. Details of Mary Ellen's story can be found in Eric A. Shelman and Stephen Lazoritz, *The Mary Ellen Wilson Child Abuse Case and the Beginning of Children's Rights in 19th Century America* (Jefferson, NC: McFarland & Company, 2005).

2. Ibid., 15.

3. Ibid., 74.

4. Ibid., 16.

5. Ibid., 8–9.

6. Ibid., 66.

7. For a review of Henry Bergh's work and influence, see Shelman and Lazoritz, *Mary Ellen Wilson Child Abuse Case*; Gerald Carson, "The Great Meddler," *American Heritage* 19 (1967), http://www.americanheritage.com/content/great-meddler?page=show; Nancy Furstinger, *Mercy: The Incredible Story of Henry Bergh, Founder of the ASPCA and Friend to Animals* (New York: Houghton Mifflin Harcourt, 2016).

8. Frans de Waal, "Are We in Anthropodenial?" *Discover*, July 1, 1997, http://discovermagazine.com/1997/jul/areweinanthropod1180.

9. Ibid.

10. *English Oxford Living Dictionaries*, s.v., "principle," https://en.oxforddictionaries.com/definition/principle.

11. Thomas L. Friedman, *Thank You for Being Late: An Optimist's Guide to Thriving in the Age of Accelerations* (New York: Farrar, Straus & Giroux, 2016).

CHAPTER 1

1. Saranga Jain and Kathleen Kurz, *New Insights on Preventing Child Marriage: A Global Analysis of Factors and Programs* (Washington, DC: International Center for Research on Women, 2007), 8, 2017, https://www.icrw.org/wp-content/uploads/2016/10/New-Insights-on-Preventing-Child-Marriage.pdf

2. William Sears, *A Cry from the Heart: The Bahá'ís in Iran* (Oxford: George Ronald, 1982), 147.

3. Wisława Szymborska, *Map: Collected and Last Poems*, trans. Clare Cavanagh and Stanislaw Baranczak (New York: Houghton Mifflin Harcourt, 2015), 260.

4. More on the Cyrus Cylinder can be found in *The Cyrus Cylinder: The King of Persia's Proclamation from Ancient Babylon*, ed. Irving Finkel (New York: I. B. Tauris & Co. Ltd., 2013).

CHAPTER 2

1. Sabrina Tonutti, "Cruelty, Children, and Animals: Historically One, Not Two, Causes," in *The Link between Animal Abuse and Human Violence*, ed. Andrew Linzey (Portland, OR: Sussex Academic Press, 2009), 95–97.

2. Ibid., 99; Shelman and Lazoritz, *Mary Ellen Wilson Child Abuse Case*.

3. Eleonora Gullone, "A Lifespan Perspective on Human Aggression and Animal Abuse," in *Animal Abuse and Human Violence*, ed. Linzey, 52.

4. John P. Clarke, *New South Wales Police Animal Cruelty Research Project* (Sydney, Australia: New South Wales Police Service, 2002).

5. Lynn Loar, " 'I'll Only Help You If You Have Two Legs,' or Why Human Service Professionals Should Pay Attention to Cases Involving Cruelty to Animals," in *Child Abuse, Domestic Violence, and Animal Abuse: Linking the Circles of Compassion for Prevention and Intervention*, ed. Frank R. Ascione and Phil Arkow (West Lafayette, IN: Purdue University Press, 1999).

6. Gullone, "A Lifespan Perspective," 52.

7. See Johan Galtung, "Violence, Peace, and Peace Research," *Journal of Peace Research* 6 (1969): 167–91; Paul Farmer, "On Suffering and Structural Violence: A View from Below," *Daedalus* 125 (1996): 261–83; Pierre Bourdieu, "Social Space and Symbolic Power," *Sociological Theory* 7 (1989): 14–25; Pierre Bourdieu, *Language and Symbolic Power*, ed. John Thompson, trans. Gino Raymond and Matthew Adamson (Cambridge, MA: Polity Press, 1991).

8. Amy J. Fitzgerald, Linda Kalof, and Thomas Dietz. "Slaughterhouse and Increased Crime Rates: An Empirical Analysis of the Spillover from 'The Jungle' Into the Surrounding Community," *Organization and Environment* 22 (2009): 158–84.

9. George Yancy and Peter Singer, "Peter Singer: On Racism, Animal Rights and Human Rights," *New York Times*, May 27, 2015, https://opinionator.blogs.nytimes.com/2015/05/27/peter-singer-on-speciesism-and-racism/.

10. Kimberly Costello and Gordon Hodson, "Explaining Dehumanization among Children: The Interspecies Model of Prejudice," *British Journal of Social Psychology* 53 (2014): 175–97.

11. For a review of connections between the treatment of women and animals, see Carol J. Adams, *The Sexual Politics of Meat: A Feminist-Vegetarian Critical Theory* (New York: Bloomsbury Academic, 2015).

12. For more on attempts to dehumanize humans based on racial classifications, including through comparisons to animals, see David Livingstone Smith, *Less Than Human: Why We Demean, Enslave, and Exterminate Others* (New York: St. Martin's Press, 2011).

13. Brian G. Henning, *The Ethics of Creativity: Beauty, Morality, and Nature in a Processive Cosmos* (Pittsburgh: University of Pittsburgh Press, 2005), 12–15.

14. See Lynne U. Sneddon, "Pain Perception in Fish: Indicators and Endpoints," *ILAR Journal* 50 (2009): 338–42; Culum Brown, "Fish Intelligence, Sentience, and Ethics," *Animal Cognition* 18 (2015): 1–17.

15. Giorgia della Rocca, Alessandra Di Salvo, Giacomo Giannettoni, and Mary Ellen Goldberg. "Pain and Suffering in Invertebrates: An Insight on Cephalopods," *American Journal of Animal and Veterinary Sciences* 10 (2015): 77–84.

16. For a review of how sickness behavior can manifest in animals, see Benjamin L. Hart, "Biological Basis of the Behavior of Sick Animals," *Neuroscience and Biobehavioral Reviews* 12 (1988): 123–37.

17. For further details on the use of animals in psychiatric research and corresponding ethical problems, see Hope Ferdowsian, "Ethical Problems Concerning the Use of Animals in Psychiatric Research," in *The Oxford Handbook of Psychiatric Ethics*, ed. John Z. Sadler, K. W. M. Fulford, and Cornelius Werendly van Staden (New York: Oxford University Press, 2015), 989–1007.

18. For a summary of Harry Harlow's experiments, see Harry Harlow, ed., *From Learning to Love: The Selected Papers of H. F. Harlow* (New York: Praeger Publishers, 1986).

19. See Ferdowsian, "Ethical Problems."

20. See, for example, Sahar Akhtar, "Animal Pain and Welfare: Can Pain Sometimes Be Worse for Them Than for Us?" in *The Oxford Handbook of Animal Ethics*, ed. Tom L. Beauchamp and R. G. Frey (New York: Oxford University Press, 2011), 495–518.

21. Hope Ferdowsian and Debra Merskin, "Parallels in Sources of Trauma, Pain, Distress, and Suffering in Humans and Nonhuman Animals," *Journal of Trauma and Dissociation* 13 (2012): 448–68.

22. Catriona Mackenzie, Wendy Rogers, and Susan Dodds, eds., *Vulnerability: New Essays in Ethics and Feminist Philosophy* (New York: Oxford University Press, 2014).

23. Jane Johnson, "Vulnerable Subjects? The Case of Nonhuman Animals in Experimentation," *Journal of Bioethical Inquiry* 10 (2013): 497–504.

24. See Thomas Nagel, "What Is It Like to Be a Bat?" *Philosophical Review* 83 (1974): 435–50.

25. Rebecca Riffkin, "In U.S., More Say Animals Should Have Same Rights as People," Gallup, May 18, 2015, http://www.gallup.com/poll/183275/say-animals-rights-people.aspx.

26. "A Science Odyssey: People and Discoveries: Abraham Maslow," PBS, accessed January 11, 2017, http://www.pbs.org/wgbh/aso/databank/entries/bhmasl.html.

27. Abraham H. Maslow, *The Journals of A. H. Maslow* (Monterey, CA: Brooks/Cole Publishing, 1979), 331.

28. Maslow describes these needs in detail in Abraham H. Maslow, "A Theory of Human Motivation," *Psychological Review* 50 (1943): 370–96.

29. Abraham H. Maslow, *The Farther Reaches of Human Nature* (New York, Viking Press, 1971), 138–40.

30. Quoted in Matthieu Ricard and Trinh Xuan Thuan, *The Quantum and the Lotus: A Journey to the Frontiers Where Science and Buddhism Meet* (New York: Three Rivers Press, 2001), 72.

31. Brené Brown, *Rising Strong: The Reckoning. The Rumble. The Revolution* (New York: Spiegel & Grau, 2015).

CHAPTER 3

1. For more on the behavioral, anatomical, and physiological similarities in how people and animals suffer, see Franklin D. McMillan, ed., *Mental Health and Well-Being in Animals* (Ames, IA: Blackwell Publishing Professional, 2005); Neville G. Gregory, *Physiology and Behaviour of Animal Suffering* (Ames, IA: Blackwell Publishing Professional, 2004).

2. See, for example, Martin Brüne, *Textbook of Evolutionary Psychiatry: The Origins of Psychopathology* (New York: Oxford University Press, 2008); Martin Brüne, Ute Brüne-Cohrs, William C. McGrew, and Signe Preuschoft, "Psychopathology in Great Apes: Concepts, Treatment Options and Possible Homologies to Human Psychiatric Disorders," *Neuroscience and Biobehavioral Reviews* 30 (2006): 1246–59; G. A. Bradshaw, Theodora Capaldo, Lorin Lindner, and Gloria Grow, "Building an Inner Sanctuary: Complex PTSD in Chimpanzees," *Journal of Trauma and Dissociation* 9 (2008): 9–34.

3. The method developed by Michael Scheeringa and his colleagues, along with their early findings, is described in the following papers: Michael S. Scheeringa, Charles H. Zeanah, Martin J. Drell, and Julie A. Larrieu, "Two Approaches to the Diagnosis of Posttraumatic Stress Disorder in Infancy and Early Childhood," *Child and Adolescent Psychiatry* 34 (2005): 191–200; Michael S. Scheeringa, Charles H. Zeanah, Leann Myers, and Frank W. Putnam, "New Findings on Alternative Criteria for PTSD in Preschool Children," *Child and Adolescent Psychiatry* 42 (2003): 561–70.

4. Roger Fouts, *Next of Kin: What Chimpanzees Have Taught Me about Who We Are* (New York: William Morrow & Company, 1997), 356; Roger S. Fouts and Deborah Fouts, "My Brother's Keeper: A Reflection on Booee," *Satya*, December 1996, http://www.satyamag.com/dec96/keeper.html; "Roger Fouts and Booee's Reunion," YouTube, published on February 23, 2014, https://www.youtube.com/watch?v=wIvmI7DxWRo.

5. The study, methods, and findings are detailed in Hope Ferdowsian, Debra Durham, Charles Kimwele, Godelieve Kranendonk, Emily Otali, Timothy Akugizibwe, J. B. Mulcahy, Lilly Ajarova, and Cassie M. Johnson, "Signs of Mood and Anxiety Disorders in Chimpanzees," *PLoS ONE* 6(6) (2011): e19855, doi:10.1371/journal.pone.00198.

6. Hope Ferdowsian, Debra Durham, Cassie M. Johnson, Martin Brüne, Charles Kimwele, Godelieve Kranendonk, Emily Otali, Timothy Akugizibwe, J. B. Mulcahy, and Lilly Ajarova, "Signs of Generalized Anxiety and Compulsive Disorders in Chimpanzees," *Journal of Veterinary Behavior: Clinical Applications and Research* 7 (2012): 353–61.

7. Sonya M. Kahlenberg and Richard W. Wrangham, "Sex Differences in Chimpanzees' Use of Sticks as Play Objects Resemble Those of Children," *Current Biology* 20 (2010): R1067–68.

8. John P. Gluck, *Voracious Science and Vulnerable Animals* (Chicago: University of Chicago Press, 2016).

9. Sue Savage-Rumbaugh, Kanzi Wamba, Panbanisha Wamba, and Nyota Wamba. "Welfare of Apes in Captive Environments: Comments on, and by, a Specific Group of Apes," *Journal of Applied Animal Welfare Science* 10 (2007): 7–19.

10. Martin Brüne, "Psychopathology in Hominoids: Do Apes Present Treatable Psychiatric Conditions?" paper presented at the annual meeting for the American Association for the Advancement of Science (AAAS), Boston, February 14–18, 2013. Also see Pallab Ghosh, "Lab Chimps Successfully Treated with Anti-Depressants," BBC News, February 14, 2013, http://www.bbc.com/news/science-environment-21299657.

11. "The Last 1,000," last update September 16, 2016, http://last1000chimps.com.

12. Bruce M. Altevogt and the Committee on the Use of Chimpanzees in Biomedical and Behavioral Research, eds., *Chimpanzees in Biomedical and Behavioral Research: Assessing the Necessity* (Washington, DC: National Academies Press, 2011).

13. See Tom L. Beauchamp, Hope R. Ferdowsian, and John Gluck, "Where Are We in the Justification of Research Involving Chimpanzees?" *Kennedy Institute of Ethics Journal* 22 (2012): 211–42.

14. See AWA, 2010, Chapter 54—Transportation, Sale, and Handling of Certain Animals, 7 U.S.C. 54, § 2131, accessed January 11, 2017, http://www.gpo.gov/fdsys/pkg/USCODE-2009-title7/html/USCODE-2009-title7-chap54.htm.

15. For a historical perspective on torture and impunity, see Alfred W. McCoy, *Torture and Impunity* (Madison: University of Wisconsin Press, 2012).

16. Amnesty International, "Report on Torture," January 1, 1973, https://www.amnesty.org/en/documents/ACT40/001/1973/en/.

17. Amnesty International, "Torture in the Eighties," January 1, 1984, https://www.amnesty.org/en/documents/ACT40/001/1984/en/.

18. *Convention against Torture and Other Cruel, Inhuman, or Degrading Treatment or Punishment*, Part 1, Article 1, accessed January 11, 2017, http://www.ohchr.org/EN/ProfessionalInterest/Pages/CAT.aspx.

19. See McCoy, *Torture and Impunity*, 151–87.

20. Senate Select Committee on Intelligence, *Committee Study of the Central Intelligence Agency's Detention and Interrogation Program*, updated for release April 3, 2014, https://web.archive.org/web/20141209165504/http://www.intelligence.senate.gov/study2014/sscistudy1.pdf, 42.

21. Adam Goldman and Kathy Gannon, "Death Shed Light on CIA 'Salt Pit' Near Kabul," Associated Press, March 28, 2010.

22. McCoy, *Torture and Impunity*, 5.

23. Steven M. Wise, "Nonhuman Rights to Personhood," *Pace Environmental Law Review* 30 (2013): 1278–90, http://digitalcommons.pace.edu/pelr/vol30/iss3/10.

24. Charles Siebert, "Should a Chimp Be Able to Sue Its Owner?" *New York Times Magazine*, April 23, 2014, https://www.nytimes.com/2014/04/27/magazine/the-rights-of-man-and-beast.html.

25. Case transcript, *In the Matter of a Proceeding under Article 70 of the CPLR for a Writ of Habeas Corpus v. Patrick C. Lavery*, quoted in Steven M. Wise, *Rattling the Cage: Toward Legal Rights for Animals* (Boston: Da Capo Press, 2014), xxiii.

26. See, for example, Wise, "Nonhuman Rights to Personhood"; Wise, *Rattling the Cage*.

27. Quoted in Wise, *Rattling the Cage*, xxiii.

28. Quoted in James C. McKinley, "Arguing in Court Whether 2 Chimps Have the Right to 'Bodily Liberty,'" *New York Times*, May 27, 2015, https://www.nytimes.com/2015/05/28/nyregion/arguing-in-court-whether-2-chimps-have-the-right-to-bodily-liberty.html.

29. Steven M. Wise, *Though the Heavens May Fall: The Landmark Trial That Led to the End of Human Slavery* (Boston: Da Capo Press, 2005).

30. Granville Sharp, *A Representation of the Injustice and Dangerous Tendency of Tolerating Slavery: Or of Admitting the Least Claim of Private Property in the Persons of Men in England* (London: Benjamin White and Robert Horsfield, 1769), https://archive.org/details/representationofooshar.

31. Prince Hoare, *Memoirs of Granville Sharp, Esq. Composed from His Own Manuscripts and Other Authentic Documents in the Possession of His Family and of the African Institution* (London: Henry Colburn and Co., 1828), 453.

CHAPTER 4

1. See Yancy and Singer, "Peter Singer"; referring to Dick Gregory, "The Circus: It's Modern Slavery," *Marin Independent Journal*, April 28, 1998.

2. For a chronology of the history of slavery and an account of modern slavery, see Kevin Bales, *New Slavery: A Reference Handbook*, 2nd ed. (Santa Barbara, CA: ABC-CLIO, 2005).

3. Charles Siebert, "An Elephant Crackup?" *New York Times Magazine*, October 8, 2006, http://www.nytimes.com/2006/10/08/magazine/08elephant.html.

4. G. A. Bradshaw, *Elephants on the Edge: What Animals Teach Us about Humanity* (New Haven, CT: Yale University Press, 2009).

5. G. A. Bradshaw and Robert M. Sapolsky. "Macroscope: Mirror, Mirror," *American Scientist* 94 (2006): 487–89.

6. G. A. Bradshaw and Lorin Lindner, "Post-Traumatic Stress and Elephants in Captivity," accessed January 11, 2017, https://pdfs.semanticscholar.org/b5b9/4307d6f45747fdccfb82dc5a69df71d02658.pdf.

7. For a tribute to Joanna Burke, see Rick Foster, "For the Love of Joanna . . . and the Elephants," *Sun Chronicle*, March 28, 2010, http://www.thesunchronicle.com/news/for-the-love-of-joanna-and-the-elephants/article_a4fa277d-9d20-54cb-9d39-f3c026d45e2c.html.

8. Ibid.

9. Kristin M. Hall, "Elephant Spared Euthanization: Handler's Death Ruled an Accident," Associated Press, July 25, 2006.

10. See Martin E. Seligman and Steven F. Maier, "Failure to Escape Traumatic Shock," *Journal of Experimental Psychology* 74 (1967): 1–9; Martin E. Seligman, "Learned Helplessness," *Annual Review of Medicine* 23 (1972): 407–12.

11. See Maria Konnikova, "Trying to Cure Depression, but Inspiring Torture," *New Yorker*, January 14, 2015, http://www.newyorker.com/science/maria-konnikova/theory-psychology-justified-torture; Terrence McCoy, "'Learned Helplessness': The Chilling Psychological Concept Behind the CIA's Interrogation Methods," *Washington Post*, December 11, 2014, https://www.washingtonpost.com/news/morning-mix/wp/2014/12/11/the-chilling-psychological-principle-behind-the-cias-interrogation-methods/?utm_term=.bae74879947b.

12. See, for example, Brunno R. Levone, John F. Cryan, and Olivia F. O'Leary, "Role of Adult Hippocampal Neurogenesis in Stress Resilience," *Neurobiology of Stress* 1 (2015): 147–55; Jeansok J. Kim, Eun Yong Song, and Therese A. Kosten, "Stress Effects in the Hippocampus: Synaptic Plasticity and Memory," *Stress* 9 (2006): 1–11; Ronald S. Duman, "Neural Plasticity: Consequences of Stress and Actions of Antidepressant Treatment," *Dialogues in Clinical Neuroscience* 6 (2004): 157–69.

13. See, for example, Tom L. Beauchamp and Victoria Wobber, "Autonomy in Chimpanzees," *Theoretical Medicine and Bioethics* 35 (2014): 117–32; Tom Regan, *The Case for Animal Rights* (Berkeley: University of California Press, 2004); Jessica Pierce, "Animals and Autonomy," *Psychology Today*, March 10, 2013, https://www.psychologytoday.com/blog/all-dogs-go-heaven/201303/animals-and-autonomy.

14. For further discussion, see Christine M. Korsgaard, "Interacting with Animals: A Kantian Account," *The Oxford Handbook of Animal Ethics*, ed. Tom L. Beauchamp and R. G. Frey (New York: Oxford University Press, 2011), 91–118. On moral judgments in animals, see Marc Bekoff and Jessica Pierce, *Wild Justice: The Moral Lives of Animals* (Chicago: University of Chicago Press, 2009).

15. Sue Donaldson and Will Kymlicka, *Zoopolis: A Political Theory of Animal Rights* (New York: Oxford University Press, 2011).

16. Muhamed Sacirbey, "Do Animals Need a UN Ambassador?" *Huffington Post*, July 4, 2014, http://www.huffingtonpost.com/ambassador-muhamed-sacirbey/do-animals-need-un-ambass_b_5558663.html.

17. Sacirbey, "Do Animals Need a UN Ambassador?"

18. "UN Court Rules against Japan's Whaling Activities in the Antarctic," UN News Centre, March 31, 2014, http://www.un.org/apps/news/story.asp?NewsID=47468#.WHp4s7HMzUY.

19. "Federal Appeals Court Rejects Navy Sonar-Use Rules," Associated Press, July 16, 2016, http://bigstory.ap.org/article/b963fd72ddc04597a558749f74b8ff73/federal-appeals-court-rejects-navy-sonar-use-rules.

CHAPTER 5

1. Carl Safina, *Beyond Words: What Animals Think and Feel* (New York: Henry Holt and Company, 2015), 151–59.

2. Safina, *Beyond Words*, 156.

3. Epidemiology Program, Post Deployment Health Group, Office of Public Health Veterans Health Administration, Department of Veterans Affairs, *Report on VA Facility Specific Operation Enduring Freedom, Operation Iraqi Freedom, and Operation New Dawn Veterans Coded with Potential PTSD, from 1st Qtr FY 2002 through 3rd Qtr FY 2012* (Washington, DC: Author, 2012), http://www.publichealth.va.gov/docs/epidemiology/ptsd-report-fy2012-qtr3.pdf.

4. See "Living with Wolves Saved My Life," Great Big Story, accessed January 17, 2017, http://www.greatbigstory.com/stories/wolf-therapy-veteran-ptsd-lockwood-animal.

5. Lorin Lindner, "To Love Like a Bird," in *Kinship with the Animals*, ed. Michael Tobias and Kate Solisti-Mattelon (Hillsboro, OR: Beyond Words Publishing, 1998), 55.

6. Lorin provides an account of her relationship with Sam and Manny in Lindner, "To Love Like a Bird," 53–63.

7. G. A. Bradshaw, Joseph P. Yenkosky, and Eileen McCarthy. "Avian Affective Dysregulation: Psychiatric Models and Treatment for Parrots in Captivity," in *Proceedings of the 30th Annual Association of Avian Veterinarians Conference*, accessed January 11, 2017, http://kerulos.org/wp-content/uploads/2013/11/Bradshaw_Yenkosky_McCarthy_910_FINAL_8.13.09_AAV-TABLES.pdf.

8. See Lorin Lindner, "The Parrots of Serenity Park," in *Kinship with Animals*, ed. Kate Solisti and Michael Tobias, updated ed. (San Francisco: Council Oak Books, 2006), 58–60.

9. See Charles Siebert, "What Does a Parrot Know about PTSD?" *New York Times Magazine*, January 28, 2016, https://www.nytimes.com/2016/01/31/magazine/what-does-a-parrot-know-about-ptsd.html?_r=0.

10. See, for example, John P. Wilson and Beverley Raphael, eds., *International Handbook of Traumatic Stress Syndromes* (New York: Springer Science+Business Media, 1993).

11. Barbara L. Fredrickson, Michele M. Tugade, Christian E. Waugh, and Gregory R. Larkin, "What Good Are Positive Emotions in Crises? A Prospective Study of Resilience and Emotions Following the Terrorist Attacks on the United States on September 11th, 2001," *Journal of Personality and Social Psychology* 84 (2003): 365–76. Also see Sandro Galea, Jennifer Ahern, Heidi Resnick, Dean Kilpatrick, Michael Bucuvalas, Joel Gold, and David Vlahov, "Psychological Sequelae of the September 11 Terrorist Attacks in New York City," *New England Journal of Medicine* 346 (2002): 982–87.

12. Glenn N. Levine, Karen Allen, Lynne T. Braun, Hayley E. Christian, Erika Friedmann, Kathryn A. Taubert, Sue Ann Thomas, Deborah L. Wells, and Richard A. Lange, on behalf of the American Heart Association Council on Clinical

Cardiology and Council on Cardiovascular and Stroke Nursing, "Pet Ownership and Cardiovascular Risk: A Scientific Statement from the American Heart Association," *Circulation* 127 (2013): 2353–63.

13. See, for example, Camelia E. Hostinar, Regina M. Sullivan, and Megan R. Gunnar, "Psychobiological Mechanisms Underlying the Social Buffering of the HPA Axis: A Review of Animal Models and Human Studies across Development," *Psychological Bulletin* 140 (2014): 256–82; Fuxia Xiong and Lubo Zhang, "Role of the Hypothalamic-Pituitary-Adrenal Axis in Developmental Programming of Health and Disease," *Frontiers in Neuroendocrinology* 34 (2013): 27–46; Nikolaos P. Daskalakis and Rachel Yehuda, "Early Maternal Influences on Stress Circuitry: Implications for Resilience and Susceptibility to Physical and Mental Disorders," *Frontiers in Endocrinology* 5 (2015): 5–6.

14. Theresa Stichick Betancourt and Kashif Tanveer Khan, "The Mental Health of Children Affected by Armed Conflict: Protective Processes and Pathways to Resilience," *International Review of Psychiatry* 20 (2008): 317–28.

15. On the benefits of empathy and altruism, see Charles Daniel Batson, *Altruism in Humans* (New York: Oxford University Press, 2011). For a partial review of the neural basis of empathy, see Boris C. Bernhardt and Tania Singer, "The Neural Basis of Empathy," *Annual Review of Neuroscience* 35 (2012): 1–23.

16. See, for example, Frans de Waal, *The Age of Empathy: Nature's Lessons for a Kinder Society* (New York: Three Rivers Press, 2009).

17. Paul J. Zak, *The Moral Molecule: The Source of Love and Prosperity* (New York: Penguin Group, 2012).

18. Office of the Child Advocate, State of Connecticut, *Shooting at Sandy Hook Elementary School*, November 21, 2014, http://www.ct.gov/oca/lib/oca/sandyhook11212014.pdf.

19. Nicole Hockley gave a keynote address about her son and the organization Sandy Hook Promise at the 19th International Conference and Summit on Violence, Abuse, and Trauma, in San Diego, CA, September 7–10, 2014.

20. Maya Angelou, *Touched by an Angel*, 1995.

CHAPTER 6

1. John Rawls, *A Theory of Justice* (Cambridge, MA: Belknap Press of Harvard University Press, 1971).

2. Rawls, *A Theory of Justice*, 136–42.

3. "Child Abuse and Neglect Fatalities 2014: Statistics and Interventions," fact sheet, accessed January 23, 2017, https://www.childwelfare.gov/pubPDFs/fatality.pdf.

4. Ellen L. Bassuk, Carmela J. DeCandia, Corey Anne Beach, and Fred Berman, *America's Youngest Outcasts: A Report Card on Child Homelessness* (Waltham, MA: National Center of Family Homelessness at American Institutes for Research, 2014), http://www.air.org/sites/default/files/downloads/report/Americas-Youngest-Outcasts-Child-Homelessness-Nov2014.pdf.

5. US Department of Health and Human Services, *Youth with Runaway, Throwaway, and Homeless Experiences: Prevalence, Drug Use, and Other At-Risk Behaviors* (Washington, DC: Author, 1995).

6. US Department of Health and Human Services, *National Evaluation of Runaway and Homeless Youth* (Washington, DC: Author, 1997).

7. Holly H. McManus and Sanna J. Thompson, "Trauma among Unaccompanied Homeless Youth: The Integration of Street Culture into a Model of Intervention," *Journal of Aggression, Maltreatment and Trauma* 16 (2008): 92–109.

8. See Colette L. Auerswald, Jessica S. Lin, and Andrea Parriott, "Six-Year Mortality in a Street-Recruited Cohort of Homeless Youth in San Francisco, California," *PeerJ*, April 14, 2016, https://peerj.com/articles/1909.pdf, doi:10.7717/peerj.1909; Pete Axthelm, "Somebody Else's Kids," *Newsweek*, April 25, 1988, 64–68; Deborah J. Sherman, "The Neglected Health Care Needs of Street Youth," *Public Health Reports* 107 (1992): 433–40.

9. This estimate is according to Pets of the Homeless (https://www.petsofthehomeless.org/about-us/faqs/). Rates are estimates and vary by region.

10. See Leslie Irvine, *My Dog Always Eats First: Homeless People and Their Animals* (Boulder, CO: Lynne Rienner Publishers, 2013).

11. Florence Nightingale, *Notes on Nursing* (Philadelphia: J. B. Lippincott, [1859] 1992).

12. Tracy J. Dietz, Diana Davis, and Jacquelyn Pennings, "Evaluating Animal-Assisted Therapy in Group Treatment for Child Sexual Abuse," *Journal of Child Sexual Abuse* 21 (2012): 665–83.

13. Nekane Balluerka, Alexander Muela, Nora Amiano, and Miguel A. Caldentey, "Promoting Psychosocial Adaptation of Youths in Residential Care Through Animal-Assisted Psychotherapy," *Child Abuse and Neglect* 50 (2015): 193–205.

14. Marc Bekoff and Jessica Pierce, *Wild Justice: The Moral Lives of Animals* (Chicago: University of Chicago Press, 2009); Jessica Pierce and Marc Bekoff, "Wild Justice Redux: What We Know about Social Justice in Animals and Why It Matters," *Social Justice Research* 25 (2012); 122–39.

15. Jason M. Cowell and Jean Decety, "Precursors to Morality in Development as a Complex Interplay between Neural, Socioenvironmental, and Behavioral Facets," *Proceedings of the National Academy of Sciences of the United States of America* 112 (2015): 12657–62.

16. Orlaith N. Fraser, Daniel Stahl, and Filippo Aureli, "Stress Reduction Through Consolation in Chimpanzees," *Proceedings of the National Academy of Sciences of the United States of America* 105 (2008): 8557–62.

17. Darby Proctor, Rebecca A. Williamson, Frans B. M. de Waal, and Sarah F. Brosnan, "Chimpanzees Play the Ultimatum Game," *Proceedings of the National Academy of Sciences of the United States of America* 110 (2013): 2070–75. Also see Sarah F. Brosnan, "Justice- and Fairness-Related Behaviors in Nonhuman Primates," *Proceedings of the National Academy of Sciences of the United States of America* 110 (2013): 10416–23.

18. Marc Bekoff, "Preface," in Lori Gruen, *Entangled Empathy: An Alternative Ethic for Our Relationships with Animals* (New York: Lantern Books, 2015), ix.

19. Bernd Heinrich, *Mind of the Raven* (New York: Cliff Street Books, 1999); Sarah Brosnan, "Nonhuman Species' Reactions to Inequity and Their Implications for Fairness," *Social Justice Research* 19 (2006): 153–85.

20. Bernd Heinrich and Thomas Bugnyar, "Just How Smart Are Ravens?" *Scientific American*, April 1, 2007, https://www.scientificamerican.com/article/just-how-smart-are-ravens/.

21. Maarten Kunst, Lieke Popelier, and Ellen Varekamp, "Victim Satisfaction with the Criminal Justice System and Emotional Recovery: A Systematic and Critical Review of the Literature," *Trauma, Violence, and Abuse* 16 (2015): 336–58.

22. Golnaz Tabibnia, Ajay B. Satpute, and Matthew D. Lieberman, "The Sunny Side of Fairness: Preference for Fairness Activates Reward Circuitry (and Disregarding Unfairness Activates Self-Control Circuitry)," *Psychological Science* 19 (2008): 339–47. Also see Alexander W. Cappelen, Tom Eichele, Kenneth Hugdahl, Karsten Specht, Erik Sorensen, and Bertil Tungodden, "Equity Theory and Fair Inequality: A Neuroeconomic Study," *Proceedings of the National Academy of Sciences of the United States of America* 111 (2014): 15368–72; Elizabeth Tricom, Antonio Rangel, Colin F. Camerer, and John P. O'Doherty, "Neural Evidence for Inequality-Averse Social Preferences," *Nature* 463 (2010): 1089–91.

23. See Sarah Spinks, "Adolescent Brains Are Works in Progress: Here's Why," *Frontline*, http://www.pbs.org/wgbh/pages/frontline/shows/teenbrain/work/adolescent.html; Jay N. Giedd, Jonathan Blumenthal, Neal O. Jeffries, F. X. Castellanos, Hong Liu, Alex Zijdenbos, Tomas Paus, Alan C. Evans, and Judith L. Rapoport, "Brain Development During Childhood and Adolescence: A Longitudinal MRI Study," *Nature Neuroscience* 2 (1999): 861–63.

24. Cowell and Decety, "Precursors to Morality."

25. Ani Satz, "Animals as Vulnerable Subjects: Beyond Interest-Convergence, Hierarchy, and Property," *Animal Law* 16 (2009): 1–50.

26. Ibid., 6.

27. See Choe Chong-Smith, "Confronting Ethical Permissibility in Animal Research: Rejecting a Common Assumption and Extending a Principle of Justice," *Theoretical Medicine and Bioethics* 35 (2014): 175–85.

28. The full transcript of his speech, delivered March 25, 1965, and titled "Our God is Marching On!," sometimes referred to as "How Long, Not Long," can be found at the *Martin Luther King, Jr. Research and Education Institute*, Stanford University, accessed January 11, 2017, https://kinginstitute.stanford.edu/our-god-marching.

CHAPTER 7

1. Amber Peterman, Tia Palermo, and Caryn Bredenkamp, "Estimates and Determinants of Sexual Violence Against Women in the Democratic Republic of Congo," *American Journal of Public Health* 101 (2011): 1060–67.

2. See, for example, Denis M. Mukwege and Cathy Nangini, "Rape with Extreme Violence: The New Pathology in South Kivu, Democratic Republic of Congo," *PLOS Medicine* 6 (2009), doi:10.1371/journal.pmed.1000204.

3. Anne Firth Murray, *The Unjust and Unhealthy Situation of Women in Poorer Countries and What They Are Doing about It*, 2nd ed. (Menlo Park, CA: Author, 2013), xvi.

4. "Welcome to Advanced Digital Changemaking!" World Pulse, accessed January 18, 2017, https://www.worldpulse.com/en/community/training/advanced/learning-materials.

5. H. Slegh, G. Barker, and R. Levtov, *Gender Relations, Sexual Violence and the Effects of Conflict on Women and Men in North Kivu, Eastern Democratic Republic of Congo: Results from the International Men and Gender Equality Survey (IMAGES)* (Washington, DC, and Capetown, South Africa: Promundo-US and Sonke Gender Justice, 2014), http://promundoglobal.org/wp-content/uploads/2014/12/Gender-Relations-Sexual-and-Gender-Based-Violence-and-the-Effects-of-Conflict-on-Women-and-Men-in-North-Kivu-Eastern-DRC-Results-from-IMAGES.pdf, 53.

6. Lauren Wolfe, "Simple Innovation Keeps Girls in School, Away from Child Marriage, in DRC," *Women Under Siege*, February 17, 2016, http://www.womenundersiegeproject.org/blog/entry/simple-innovation-keeps-girls-in-school-away-from-child-marriage-in-drc.

7. Jonny Hogg, "Congo Rebels Seize Eastern City as UN Forces Look On," Reuters, November 20, 2012, http://www.reuters.com/article/us-congo-democratic-idUSBRE8AI0UO20121120.

8. Neema Namadamu, "From the Grassroots Women Leaders of Congo to the Women Leaders of the White House," Change.org petition, accessed January 11, 2017, https://www.change.org/p/from-the-grassroots-women-leaders-of-congo-to-the-women-leaders-of-the-white-house.

9. Stuart A. Reid, "Did Russ Feingold Just End a War?" *POLITICO Magazine*, March 11, 2014, http://www.politico.com/magazine/story/2014/03/russ-feingold-congo-104535_full.html#.WHuvlLHMzUY.

10. See Pete Kowalczyk, "These Park Rangers Would Die (and Have) to Save Mountain Gorillas," CNN, November 25, 2015, http://www.cnn.com/2015/11/12/africa/virunga-national-park-mountain-gorilla/; Daniel Wesangula, "Virunga Ranger Killed by Mai Mai Rebels While Protecting Gorillas," *Guardian*, December 16, 2016, https://www.theguardian.com/environment/2016/dec/16/virunga-ranger-killed-rebels-protecting-gorillas.

11. Mark Jenkins, "Who Murdered the Virunga Gorillas?" *National Geographic*, July 2008, http://ngm.nationalgeographic.com/2008/07/virunga/jenkins-text.

12. Ibid.

13. Alistair Thomson, "Congo Arrests Ranger over Gorilla Killings," Reuters, March 19, 2008, http://www.reuters.com/article/us-congo-democratic-gorillas-idUSL1973887820080319; Daniel Howden, "Gorilla Warfare: The Battle to Save One of Africa's Rarest Animals," *Independent*, October 16, 2009, http://www.independent

.co.uk/environment/nature/gorilla-warfare-the-battle-to-save-one-of-africas-rarest-animals-1803193.html.

14. Simon Worrall, "A Prince Battles to Save Gorillas amid Brutal Conflict," *National Geographic,* June 11, 2015, http://news.nationalgeographic.com/2015/06/150611-virunga-national-park-emmanuel-de-merode-africa-world/.

15. Jessica Hatcher, "Meet the First Female Rangers to Guard One of World's Deadliest Parks," *National Geographic,* October 14, 2015, http://news.nationalgeographic.com/2015/10/151014-virunga-women-rangers-mountain-gorillas-congo/.

16. Kowalczyk, "These Park Rangers."

17. Joanna Natasegara, "Wildlife Tourism in Virunga Gives New Hope to Congo," *Guardian,* November 29, 2014, https://www.theguardian.com/travel/2014/nov/29/virunga-national-park-congo. Also see "Virunga Alliance," accessed January 11, 2017, https://virunga.org/virunga-alliance/.

18. Jerome Groopman, *The Anatomy of Hope* (New York: Random House, 2005), 161.

19. See Ciro Conversano, Alessandro Rotondo, Elena Lensi, Olivia Della Vista, Francesca Arpone, and Mario Antonio Reda, "Optimism and Its Impact on Mental and Physical Well-Being," *Clinical Practice and Epidemiology in Mental Health* 6 (2010): 25–29; Tali Sharot, "The Optimism Bias," *Current Biology* 21 (2011): R941–45.

20. See, for example, Conversano et al., "Optimism and Its Impact"; Sharot, "The Optimism Bias"; C. R. Snyder, Kevin L. Rand, and David R. Sigmon, "Hope Theory: A Member of the Positive Psychology Family," in *Handbook of Positive Psychology,* ed. C. R. Snyder and Shane J. Lopez (New York: Oxford University Press, 2002), 257–76.

21. Groopman, *The Anatomy of Hope,* 172.

22. Ibid., 188.

23. Franklin D. McMillan, "The Placebo Effect in Animals," *Journal of the American Veterinary Medical Association* 215 (1999): 992–99.

24. See, for example, Michael Mendl, Oliver H. P. Burman, Richard M. A. Parker, and Elizabeth S. Paul, "Cognitive Bias as an Indicator of Animal Emotion and Welfare: Emerging Evidence and Underlying Mechanisms," *Applied Animal Behaviour Science* 118 (2009): 161–81.

25. Ibid.

CHAPTER 8

1. Charles Patterson, *Eternal Treblinka: Our Treatment of Animals and the Holocaust* (New York: Lantern Books, 2002).

2. Ibid., 53–79.

3. See, for example, Jim Mason, *An Unnatural Order: Uncovering the Roots of Our Domination of Nature and Each Other* (New York: Simon and Schuster, 1993); David Sztybel, "Can the Treatment of Animals Be Compared to the Holocaust?" *Ethics and the Environment* 11 (2006): 97–132; Patterson, *Eternal Treblinka,* 139–67.

4. See, for example, Patterson, *Eternal Treblinka,* 3–50, 81–108.

5. Upton Sinclair, *The Jungle* (New York: New American Library, [1905] 1906), 40.

6. For further reading on industrial farming in the United States, see Jonathan Safran Foer, *Eating Animals* (New York: Little, Brown and Company, 2009).

7. Patterson, *Eternal Treblinka*, 11.

8. Ibid., 11; McCoy, *Torture and Impunity*, 5.

9. Patterson, *Eternal Treblinka*, 6–12. Also see Mason, *An Unnatural Order*.

10. Patterson, *Eternal Treblinka*, 21. Also see Matthew Scully, *Dominion: The Power of Man, the Suffering of Animals, and the Call to Mercy* (New York: St. Martin's Griffin, 2002).

11. Patterson, *Eternal Treblinka*, 11.

12. Ibid., 25–26.

13. Ibid., 27–50.

14. Ibid., 49.

15. Laura Hillenbrand, *Unbroken: The World War II Story of Survival, Resilience, and Redemption* (New York: Random House, 2010), 182–83.

16. Thomas T. Hills and Stephen Butterfill, "From Foraging to Autonoetic Consciousness: The Primal Self as a Consequence of Embodied Prospective Foraging," *Current Zoology* 61 (2015): 368–81.

17. See, for example, Keith M. Kendrick, Ana P. da Costa, Andrea E. Leigh, Michael R. Hinton, and Jon W. Peirce, "Sheep Don't Forget a Face," *Nature* 414 (2007): 165–66.

18. For more on the capacities of goats, see, for example, Elodie F. Briefer, Samaah Haque, Luigi Baciadonna, and Alan G. McElligott, "Goats Excel at Learning and Remembering a Highly Novel Cognitive Task," *Frontiers in Zoology* 11 (March 26, 2014), http://frontiersinzoology.biomedcentral.com/articles/10.1186/1742-9994-11-20; Christian Nawroth, Jemma M. Brett, and Alan G. McElligott, "Goats Display Audience-Dependent Human-Directed Gazing Behaviour in a Problem-Solving Task," *Biology Letters* 12 (2016), doi:10.1098/rsbl.2016.0283; Christian Nawroth, Luigi Baciadonna, and Alan G. McElligott, "Goats Learn Socially from Humans in a Spatial Problem-Solving Task," *Animal Behaviour* 121 (2016): 123–29.

19. For a review of the intelligence of chickens, see Carolynn L. Smith and Sarah L. Zielinski, "The Startling Intelligence of the Common Chicken," *Scientific American*, February 1, 2014, https://www.scientificamerican.com/article/the-startling-intelligence-of-the-common-chicken/; Lori Marino, "Thinking Chickens: A Review of Cognition, Emotion, and Behavior in the Domestic Chicken," *Animal Cognition* 20 (2017): 127–47.

20. See, for example, Amy Hatkoff, *The Inner World of Farm Animals: Their Amazing Social, Emotional, and Intellectual Capacities* (New York: Stewart, Tabori, and Chang, 2009), 64; Kristin Hagen and Donald M. Broom, "Emotional Reactions to Learning in Cattle," *Applied Animal Behaviour Science* 85 (2004): 203–13.

21. For a review of intelligence in pigs, see Lori Marino and Christina Colvin, "Thinking Pigs: A Comparative Review of Cognition, Emotion, and Personality

in *Sus domesticus*," *International Journal of Comparative Psychology* 28 (2015), https://escholarship.org/uc/item/8sx4s79c.

22. Susie Coston, "Animals of Farm Sanctuary," July 11, 2016, Animals of Farm Sanctuary, http://www.animalsoffarmsanctuary.com/post/147257352621/remembering-rose-pig-for-whom-moving-past-trauma.

23. Ibid.

24. See Jeremy Waldron, "Dignity, Rank, and Rights," *The Tanner Lectures on Human Values*, delivered at University of California–Berkeley, April 21–23, 2009, http://tannerlectures.utah.edu/_documents/a-to-z/w/Waldron_09.pdf.

25. Ibid.

26. James Rachels, *Created from Animals: The Moral Implications of Darwinism* (New York: Oxford University Press, 1990).

27. Martha Nussbaum, *Frontiers of Justice: Disability, Nationality, Species Membership* (Cambridge, MA: Belknap Press of Harvard University Press, 2006), 346–52, 365, 375, 392–401.

28. Mary Catherine Beach, Jeremy Sugarman, Rachel L. Johnson, Jose J. Arbelaez, Patrick S. Duggan, and Lisa A. Cooper, "Do Patients Treated with Dignity Report Higher Satisfaction, Adherence, and Receipt of Preventive Care?" *Annals of Family Medicine* 3 (2005): 331–38.

29. Harvey M. Chochinov, "Dignity-Conserving Care—A New Model for Palliative Care: Helping the Patient Feel Valued," *JAMA* 1 (2002): 2253–60; Harvey M. Chochinov, Thomas Hack, Thomas Hassard, Linda J. Kristjanson, Susan McClement, and Mike Harlos, "Dignity Therapy: A Novel Psychotherapeutic Intervention for Patients Near the End of Life," *Journal of Clinical Oncology* 23 (2005): 5520–25.

30. Alix Spiegel, "For the Dying, a Chance to Rewrite Life," NPR, September 12, 2011, http://www.npr.org/2011/09/12/140336146/for-the-dying-a-chance-to-rewrite-life.

31. George Fitchett, Linda Emanuel, George Handzo, Lara Boyken, and Diana J. Wilkie, "Care of the Human Spirit and the Role of Dignity Therapy: A Systematic Review of Dignity Therapy Research," *BMC Palliative Care* 14 (2005), doi:10.1186/s12904-015-0007-1, http://bmcpalliatcare.biomedcentral.com/articles/10.1186/s12904-015-0007-1.

32. As quoted in Patterson, *Eternal Treblinka*, 183. The original story is found in "The Letter Writer" in *The Seance and Other Stories* (New York: Farrar, Straus & Giroux, 1968) and *The Collected Stories* (New York: Farrar, Straus & Giroux, 1982).

33. See Patterson, *Eternal Treblinka*, 169–200.

34. See Dale J. Langford, Sara E. Crager, Zarrar Shehzad, Shad B. Smith, Susana G. Sotocinal, Jeremy S. Levenstadt, Mona Lisa Chanda, Daniel J. Levitin, and Jeffrey S. Mogil, "Social Modulation of Pain as Evidence for Empathy in Mice," *Science* 30 (2006): 1967–970; Inbal Ben-Ami Bartal, Jean Decety, and Peggy Mason, "Empathy and Pro-Social Behavior in Rats," *Science* 334 (2011): 1427–30; Alka Chandna, "Rats Have Empathy, but What about the Scientists Who Experiment

on Them?" Hastings Center, June 24, 2015, http://www.thehastingscenter.org/rats-have-empathy-but-what-about-the-scientists-who-experiment-on-them/.

CHAPTER 9

1. He describes his interactions with Robert Lewis in John Gluck, *Voracious Science and Vulnerable Animals* (Chicago: University of Chicago Press, 2016), 286–89.

2. Roel B. van den Broek, "The Death and Rebirth of the Phoenix," in *Myth of the Phoenix According to Classical and Early Christian Traditions* (Leiden, The Netherlands: Brill Archive, 1972), 146–232.

3. Laura H. Kahn, Bruce Kaplan, Thomas P. Monath, Jack Woodall, and Lisa A. Conti, "History of the One Health Initiative Team (April 2006 through September 2015) and the One Health Initiative Website (October 2008 through September 2015)," One Health Initiative, September 25, 2015, http://www.onehealthinitiative.com/news.php?query=History+of+the+One+Health+Initiative+team+(April+2006+through+September+2015)+and+the+One+Health+Initiative+website+since+October+1,+2008.

4. The mission of the One Health Initiative is as follows: "Recognizing that human health (including mental health via the human-animal bond phenomenon), animal health, and ecosystem health are inextricably linked, One Health seeks to promote, improve, and defend the health and wellbeing of all species by enhancing cooperation and collaboration between physicians, veterinarians, other scientific health and environmental professionals and by promoting strengths in leadership and management to achieve these goals." See "Mission Statement," One Health Initiative, accessed January 11, 2017, http://www.onehealthinitiative.com/mission.php.

5. See Hope Ferdowsian, "Africa's Meat and Dairy Industry: A Threat to the Continent's Future," in *Africa and Her Animals: A Philosophical and Practical Reader*, ed. Rainer Ebert and Anteneh Roba (Pretoria: University of South Africa [Unisa] Press, forthcoming); Hope Ferdowsian, "Can Plants Save the World?" in *ReThink Food: 100+ Doctors Can't Be Wrong*, ed. Shoshana Castle and Amy-Lee Goodman (Houston: Two Skirts Productions, 2014), 377–79; Aysha Akhtar, Michael Greger, Hope Ferdowsian, and Erika Frank, "Health Professionals' Roles in Animal Agriculture, Climate Change, and Human Health," *American Journal of Preventive Medicine* 36 (2009): 182–87.

6. See Michael Greger, *Bird Flu: A Virus of Our Own Hatching* (New York: Lantern Books, 2006).

7. Robert Goodland and Jeff Anhang, "Livestock and Climate Change: What If the Key Actors in Climate Change Are . . . Cows, Pigs, and Chickens?" *World Watch Magazine* 22 (2009): 10–19, http://www.worldwatch.org/files/pdf/Livestock%20and%20Climate%20Change.pdf.

8. See Antonella Rossati, "Global Warming and Its Health Impact," *International Journal of Occupational and Environmental Medicine* 8 (2017): 7–20.

9. David Benatar, "The Chickens Come Home to Roost," *American Journal of Public Health* 97 (2007): 1545–46.

10. Ibid., 1545–46.

11. Michael Lewis, *The Undoing Project: A Friendship That Changed Our Minds* (New York: W. W. Norton & Company, 2017), 312.

12. Ibid.

13. Ibid., 303–12.

14. As Genovese was killed outside her home, some neighbors heard and ignored her cries. Since her death, the bystander effect has become a widely recognized form of apathy, a failure of empathy. Though the *New York Times* originally reported that thirty-seven witnesses were fully aware and unresponsive to the crime, it later corrected the exaggerated portrayal. See Martin Gansberg, "37 Who Saw Murder Didn't Call the Police," *New York Times*, March 27, 1964, http://www.nytimes.com/1964/03/27/37-who-saw-murder-didnt-call-the-police.html; Jim Rasenberger, "Kitty, 40 Years Later," *New York Times*, February 8, 2004, http://www.nytimes.com/2004/02/08/nyregion/kitty-40-years-later.html; Robert D. McFadden, "Winston Moseley, Who Killed Kitty Genovese, Dies in Prison at 81," *New York Times*, April 4, 2016, https://www.nytimes.com/2016/04/05/nyregion/winston-moseley-81-killer-of-kitty-genovese-dies-in-prison.html; Clyde Haberman, "Remembering Kitty Genovese," *New York Times*, April 10, 2016, https://www.nytimes.com/2016/04/11/us/remembering-kitty-genovese.html.

15. See Jeneen Interlandi, "The Brain's Empathy Gap," *New York Times Magazine*, March 19, 2015, https://www.nytimes.com/2015/03/22/magazine/the-brains-empathy-gap.html.

16. See "Will the Haunting Image of an Injured Syrian Boy Make a Difference?" transcript, *PBS News Hour*, August 18, 2016, http://www.pbs.org/newshour/bb/will-haunting-image-injured-syrian-boy-make-difference/.

17. See Patrick Kingsley, "The Death of Alan Kurdi: One Year On, Compassion Towards Refugees Fades," *Guardian*, September 2, 2016, https://www.theguardian.com/world/2016/sep/01/alan-kurdi-death-one-year-on-compassion-towards-refugees-fades; Anne Barnard, "How Omran Daqneesh, 5, Became a Symbol of Aleppo's Suffering," *New York Times*, August 18, 2016, https://www.nytimes.com/2016/08/19/world/middleeast/omran-daqneesh-syria-aleppo.html?_r=0.

18. See Brandon Ellington Patterson, "Hate Crimes Are Rising but Don't Expect Them to Be Prosecuted," *Mother Jones*, November 25, 2016, http://www.motherjones.com/politics/2016/11/heres-why-hate-crimes-are-so-hard-prosecute; Harriet Agerholm, "Brexit: Wave of Hate Crime and Racial Abuse Following EU Referendum," *Independent*, June 26, 2016, http://www.independent.co.uk/news/uk/home-news/brexit-eu-referendum-racial-racism-abuse-hate-crime-reported-latest-leave-immigration-a7104191.html; Vikram Dodd, "Police Blame Worst Rise in Recorded Hate Crime on EU Referendum," *Guardian*, July 11, 2016, https://www.theguardian.com/society/2016/jul/11/police-blame-worst-rise-in-recorded-hate-on-eu-referendum.

19. See *After Election Day: The Trump Effect: The Impact of the 2016 Presidential Election on Our Nation's Schools*, report by the Southern Poverty Law Center, November 28, 2016, https://www.splcenter.org/sites/default/files/the_trump_effect.pdf.

20. See, for example, David A. Graham, "The Empathy Gap between Paris and Beirut," *Atlantic*, November 16, 2015, http://www.theatlantic.com/international/archive/2015/11/paris-beirut-terrorism-empathy-gap/416121/; Matt Schiavenza, "Nigeria's Horror in Paris's Shadow," *Atlantic*, January 11, 2015, http://www.theatlantic.com/international/archive/2015/01/boko-harams-quiet-destruction-of-northeast-nigeria/384416/.

21. See, for example, Jason Silverstein, "I Don't Feel Your Pain," *Slate*, June 27, 2013, http://www.slate.com/articles/health_and_science/science/2013/06/racial_empathy_gap_people_don_t_perceive_pain_in_other_races.html; Interlandi, "The Brain's Empathy Gap."

22. Robert Pearl, "Why Health Care is Different If You're Black, Latino, or Poor," *Forbes*, March 5, 2015, http://www.forbes.com/sites/robertpearl/2015/03/05/healthcare-black-latino-poor/#4fa8b5e41ca7; "What We're Learning: Reducing Disparities in the Quality of Care for Racial and Ethnic Minorities Improves Care," *Quality Field Notes* 4 (Robert Wood Johnson Foundation, 2014), http://www.rwjf.org/content/dam/farm/reports/issue_briefs/2014/rwjf412949.

23. See Kelly M. Hoffman, Sophie Trawalter, Jordan R. Axt, and M. Norman Oliver, "Racial Bias in Pain Assessment and Treatment Recommendations, and False Beliefs about Biological Differences between Blacks and Whites," *Proceedings of the National Academy of Sciences* 113 (2016): 4296–4301; Matteo Forgiarini, Marcello Gallucci, and Angelo Maravita, "Racism and the Empathy for Pain on Our Skin," *Frontiers of Psychology* 2 (2011), doi:10.3389/fpsyg.2011.00108, https://www.ncbi.nlm.nih.gov/pmc/articles/PMC3108582/pdf/fpsyg-02-00108.pdf.

24. See, for example, Dylan Matthews, "Harambe the Gorilla Reveals Americans' Hypocrisy about Animal Suffering," *Vox*, May 31, 2016, http://www.vox.com/2016/5/31/11820880/harambe-gorilla-chicken-meat.

25. See, for example, Brittney Cooper, "The Conversation about Harambe Has Racist Undertones We Can't Ignore," *Cosmopolitan*, June 1, 2016, http://www.cosmopolitan.com/politics/news/a59261/harambe-gorilla-michelle-gregg/.

26. Interlandi, "The Brain's Empathy Gap."

27. Ibid.; Emile G. Bruneau and Rebecca Saxe, "Attitudes Towards the Outgroup Are Predicted by Activity in the Precuneus in Arabs and Israelis," *Neuroimage* 52 (2010): 1704–11.

28. Michele Borba, *UnSelfie: Why Empathetic Kids Succeed in Our All-about-Me World* (New York: Touchstone, 2016).

29. For an account of the association between moral identity and response during genocide, see Kristen Renwick Monroe, *Ethics in an Age of Terror and Genocide: Identity and Moral Choice* (Princeton, NJ: Princeton University Press, 2012).

30. Sujatha Ramakrishna, *Raising Kids Who Love Animals* (Author, 2012).

31. See, for example, Tom L. Beauchamp, "Rights Theory and Animal Rights," in Tom L. Beauchamp and R. G. Frey, eds., *The Oxford Handbook of Animal Ethics* (New York: Oxford University Press, 2011), 198–227.

32. Moore's law is based on an observation that the speed and power of computational processing power has doubled roughly every year since its invention. See Friedman, *Thank You for Being Late*, 19–35, 37–44.

33. Ibid., 120–21.

34. Ibid., 164–67.

35. "Floor Statement by Senator John McCain on Senate Intelligence Committee Report on CIA Interrogation Methods," December 9, 2014, http://www.mccain.senate.gov/public/index.cfm/2014/12/floor-statement-by-sen-mccain-on-senate-intelligence-committee-report-on-cia-interrogation-methods.

36. See, for example, Arie W. Kruglanski, Michele J. Gelfand, Jocelyn J. Belanger, Anna Sheveland, Malkanthi Hetiarachchi, and Rohan Gunaratna, "The Psychology of Radicalization and Deradicalization: How Significance Quest Impacts Violent Extremism," *Political Psychology* 35 (2014): 69–93; Steven Windisch, Pete Simi, Gina Sott Ligon, and Hillary McNeel, "Disengagement from Ideologically-Based and Violent Organizations: A Systematic Review of the Literature," *Journal for Deradicalization*, no. 9 (2016–17), http://journals.sfu.ca/jd/index.php/jd/article/view/72/65; Daniel Koehler, *Understanding Deradicalization: Methods, Tools and Programs for Countering Violent Extremism* (New York: Routledge, 2017).

CHAPTER 10

1. John Gray, "Steven Pinker Is Wrong about Violence and War," *Guardian*, March 13, 2015, https://www.theguardian.com/books/2015/mar/13/john-gray-steven-pinker-wrong-violence-war-declining.

2. Steven Pinker, *The Better Angels of Our Nature: Why Violence Has Declined* (New York: Penguin Books, 2011).

3. Ibid., 59–128.

4. Ibid., 571–696.

5. Ibid., 378–82.

6. See Eve Browning Cole, "Theophrastus and Aristotle on Animal Intelligence," in *Theophrastus: His Psychological, Doxographical, and Scientific Writings*, ed. William W. Fortenbaugh and Dimitri Gutas (New Brunswick, NJ: Transaction Publishers, 1992), 44–62; Andrew Linzey, "Theophrastus," in *Encyclopedia of Animal Rights and Animal Welfare*, ed. Marc Bekoff and Carron Meaney, 333–34.

7. Porphyry, *On Abstinence from Animal Food*, ed. Esme Wynne-Tyson (New York: Barnes & Noble, 1965). Also see Richard Sorabji, *Animal Minds and Human Morals: The Origins of the Western Debate* (Ithaca, NY: Cornell University Press, 1993).

8. Elizabeth Kolbert, *The Sixth Extinction: An Unnatural History* (New York: Henry Holt and Company, 2014).

9. Gray, "Steven Pinker Is Wrong."

10. For a review of this subject, see Michelle Alexander, *The New Jim Crow: Mass Incarceration in the Age of Colorblindness* (New York: New Press, 2010).

11. Ranit Mishori, Alisse Hannaford, Imran Mujawar, Hope Ferdowsian, and Sarah Kureshi, " 'Their Stories Have Changed My Life': Clinicians' Reflections on Their Experience with and Their Motivation to Conduct Asylum Evaluations," *Journal of Immigrant and Minority Health* 18 (2016): 215.

12. Mishori et al., " 'Their Stories Have Changed My Life,' " 210–18.

13. Taime Bryant, "Trauma, Law, and Advocacy for Animals," *Journal of Animal Law and Ethics* 1 (2006): 63–138.

14. Mishori et al., " 'Their Stories Have Changed My Life,' " 215.

15. Ibid., 214.

16. Amy Joscelyne, Sarah Knuckey, Margaret L. Satterthwaite, Richard A. Bryant, Meng Li, Meng Qian, and Adam D. Brown, "Mental Health Functioning in the Human Rights Field: Findings from an International Internet-Based Study," *PLoS ONE* 10 (2015), doi:10.1371/journal.pone.0145188.

17. Bryant, "Trauma, Law, and Advocacy for Animals," 114. Italics per original author.

18. See, for example, A. Breeze Harper, "Social Justice Beliefs and Addiction to Uncompassionate Consumption: Food for Thought," in *Sistah Vegan*, ed. A. Breeze Harper (New York: Lantern Books, 2010), 20–41.

19. Ibid.

20. See Zoe Weil, *The World Becomes What We Teach: Educating a Generation of Solutionaries* (New York: Lantern Books, 2016).

21. Borba, *UnSelfie*, xv.

22. Ibid., 3–7.

23. See Jessica Alexander, "Teaching Kids Empathy: In Danish Schools, It's . . . Well, It's a Piece of Cake," *Salon*, August 9, 2016, http://www.salon.com/2016/08/09/teaching-kids-empathy-in-danish-schools-its-well-its-a-piece-of-cake/.

24. Wayne Pacelle, *The Humane Economy: How Innovators and Enlightened Consumers Are Transforming the Lives of Animals* (New York: William Morrow, 2016).

25. See, for example, Debra Merskin, *Media, Minorities, and Meaning: A Critical Introduction* (New York: Peter Lang, 2010).

26. Authors and scholars Debra Merskin and Carrie Freeman have developed guidelines for media producers and consumers that address this problem. See "Animals and Media: A Style Guide for Giving Voice to the Voiceless," Animals and Media, accessed January 11, 2017, http://www.animalsandmedia.org/main/.

27. See "The Link," National Link Coalition, accessed January 11, 2017, http://nationallinkcoalition.org.

28. See "Tracking Animal Cruelty: FBI Collecting Data on Crimes against Animals," FBI, February 1, 2016, https://www.fbi.gov/news/stories/-tracking-animal-cruelty; Colby Itkowitz, "A Big Win for Animals: The FBI Now Tracks Animal Abuse Like

It Tracks Homicides," *Washington Post*, January 6, 2016, https://www.washingtonpost.com/news/inspired-life/wp/2016/01/06/a-big-win-for-animals-the-fbi-now-tracks-animal-abuse-like-it-tracks-homicides/?utm_term=.38e78f9380c8.

29. See, for example, Liana Sun Wyler and [name redacted], "International Illegal Trade in Wildlife: Threats and U.S. Policy," *Congressional Research Service Report for Congress*, July 23, 2013, https://www.everycrsreport.com/files/20130723_RL34395_a407792bf33debc34c6516538db405e8eed1e6b3.pdf; Ted Poe, "How Poaching Fuels Terrorism Funding," CNN, October 22, 2014, http://www.cnn.com/2014/10/22/opinion/poe-poaching-terrorism-funding/; Brian Clark Howard, "U.S. Steps Up Fight against Poaching and Wildlife Trafficking," *National Geographic*, July 15, 2015, http://news.nationalgeographic.com/2015/07/150715-poaching-wildlife-trafficking-elephants-rhinos-ivory/.

30. Borba introduces the idea of "Upstanders" in Borba, *UnSelfie*, 171–74.

31. See James Gleick, *Chaos: Making a New Science* (New York: Penguin Books, 2008), 11–31.

32. Ibid., 23.

INDEX